Name ______________________________ Class ______________ Date ____________

Skills Worksheet

Concept Review

MATCHING

In the space provided, write the letter of the description that best matches the term or phrase.

______ **1.** interaction between two species in which both are harmed

______ **2.** the functional role of a species within an ecosystem

______ **3.** one of the three main properties of a population

______ **4.** development of adaptations as a result of symbiotic relationships

______ **5.** maximum population that an ecosystem can support indefinitely

______ **6.** close interaction between two species in which one organism benefits while the other organism is harmed

______ **7.** the ratio of births to deaths in a population

______ **8.** maximum number of offspring that each member of a population can produce

______ **9.** a reduction in population size caused by a natural disaster

______**10.** the location where an organism lives

a. density
b. growth rate
c. reproductive potential
d. carrying capacity
e. density independent regulation
f. niche
g. habitat
h. competition
i. parasitism
j. coevolution

AF478712

MULTIPLE CHOICE

In the space provided, write the letter of the term or phrase that best completes each statement or best answers each question.

______**11.** A territory is
a. a place where one animal lives.
b. a place where people eat.
c. an area defended by one or more individuals.
d. a place for sleeping.

______**12.** Which of the following is an example of a parasite?
a. worm in your intestine
b. a lion hunting zebras
c. bee stinger in your arm
d. honeybee on a flower

_______ **13.** Bacteria in your intestines are an example of mutualism if they
- **a.** make you sick.
- **b.** have no effect on you.
- **c.** are destroyed by digestive juices.
- **d.** help you break down food.

_______ **14.** Predators ____________ kill their prey.
- **a.** always
- **b.** usually
- **c.** never
- **d.** try not to

_______ **15.** What property of a population may be described as even, clumped, or random?
- **a.** dispersion
- **b.** density
- **c.** size
- **d.** growth rate

_______ **16.** What can occur if a population has plenty of food and space, and has no competition or predators?
- **a.** reduction of carrying capacity
- **b.** exponential growth
- **c.** zero population growth
- **d.** coevolution

_______ **17.** A grizzly bear can be all of the following *except* a
- **a.** parasite.
- **b.** competitor.
- **c.** mutualist.
- **d.** predator.

_______ **18.** The "co-" in coevolution means
- **a.** apart.
- **b.** together.
- **c.** two.
- **d.** predator-prey.

_______ **19.** Which of the following has the greatest effect on reproductive potential?
- **a.** producing more offspring at a time
- **b.** reproducing more often
- **c.** having a longer life span
- **d.** reproducing earlier in life

_______ **20.** Members of a species may compete with one another for
- **a.** running faster.
- **b.** social dominance.
- **c.** giving birth.
- **d.** mutualism.

_______ **21.** A robin that does not affect the tree in which it nests is an example of
- **a.** parasitism.
- **b.** commensalism.
- **c.** mutualism.
- **d.** predation.

_______ **22.** Two species can be indirect competitors for food if they
- **a.** use the same food source at different times.
- **b.** have different food sources.
- **c.** fight over food.
- **d.** eat together peacefully.

Name ______________________________ Class _______________ Date _____________

Skills Worksheet

Critical Thinking

ANALOGIES

In the space provided, write the letter of the pair of terms or phrases that best complete the analogy shown. An analogy is a relationship between two pairs of words or phrases written as a : b :: c : d. The symbol : is read "is to," and the symbol :: is read "as."

______ **1.** carrying capacity : population size ::
- **a.** niche : habitat
- **b.** amount of water : plant growth
- **c.** death rate : birth rate
- **d.** severe weather : density-dependent deaths

______ **2.** predator : prey ::
- **a.** competition : species
- **b.** grazing : herbivores
- **c.** ants : acacia trees
- **d.** parasite : host

______ **3.** species : population ::
- **a.** heart : body
- **b.** plants : animals
- **c.** cows : herd
- **d.** sunlight : trees

______ **4.** limiting resource : carrying capacity ::
- **a.** turtle : pond
- **b.** sunlight : plant growth
- **c.** territory : density-independent deaths
- **d.** population growth : parasitism

______ **5.** density : area ::
- **a.** dispersion : niche
- **b.** leaves : forest
- **c.** growth rate : time
- **d.** habitat : niche

______ **6.** relationship : symbiosis ::
- **a.** evolution : population
- **b.** business : partnership
- **c.** mutualism : competition
- **d.** health : illness

______ **7.** births : positive growth rate ::
- **a.** reproduction : extinction
- **b.** deaths : negative growth rate
- **c.** limited resource : exponential growth
- **d.** niche : habitat

______ **8.** long generation time : short generation time ::
- **a.** dogs : cats
- **b.** ants : dogs
- **c.** elephants : bacteria
- **d.** daisies : trees

INTERPRETING OBSERVATIONS

Read the following, and answer the questions below.

Imagine that two species of monkeys are introduced to an island that provides them with an ideal habitat. One species is arboreal and eats fruits and leaves; the other is terrestrial and relies on fallen fruits and a few small insects it can pick from the ground for survival. The monkeys have an abundance of food, no local competition for the food, and no predators. After a decade, the number of frugivorous and leaf-eating arboreal monkeys increased faster than the terrestrial fruit and insect eaters.

After 20 years, the number of terrestrial monkeys in the island started to decrease rapidly, although the arboreal monkey population remained fairly stable.

9. Explain what probably caused the changes in the two monkey populations.

____

____

____

____

____

____

10. What type of biological interaction best explains the relative greater success of the arboreal monkeys?

____

____

____

____

11. Is the terrestrial population of monkeys destined to extinction? Explain.

____

____

____

____

____

AGREE OR DISAGREE

Agree or disagree with the following statements, and support your answers.

12. An ecosystem can be viewed as a host that is parasitized by the organisms that live in or on it.

13. The only competitors that humans have for food are other humans and insects.

14. The interaction generated by human intervention to protect the gray wolf population in the northwestern United States can be defined as commensalism.

REFINING CONCEPTS

The statements below challenge you to refine your understanding of concepts covered in the chapter. Think carefully, and answer the questions that follow.

15. Although there are many predators on the African savanna, none plays exactly the same role as the lion. Can any two species occupy exactly the same niche? Why or why not?

____

____

____

____

____

16. A biologist thinks that over time a parasite can influence the evolution of its host species. Do you think that she is right? Justify your answer.

____

____

____

____

____

____

____

____

17. To be considered part of the same population organisms must have a reasonable chance of mating with each other. Are two wild roses separated by a wide road part of the same population? Defend your answer.

____

____

____

____

____

____

Name ______________________________ Class _______________ Date _____________

Skills Worksheet

Active Reading

Section: How Populations Change in Size

Read the passage below and answer the questions that follow.

Over time, the growth rates of populations change because birth rates and death rates increase or decrease. Growth rates can be positive, negative, or zero. For a population's growth rate to be zero, the average number of births must equal the average number of deaths. A population would remain the same size if each pair of adults produced exactly two offspring, and each of those offspring survived to reproduce. If the adults in a population are not replaced by new births, the growth rate will be negative and the population will shrink.

IDENTIFYING MAIN IDEAS

One reading skill is the ability to identify the main idea of a passage. The main idea is the main focus or key idea. Frequently, a main idea is accompanied by supporting information that offers detailed facts about main ideas.

In the space provided, write the letter of the term or phrase that best matches the description.

______ **1.** The average number of deaths is greater than the average number of births.

______ **2.** The average number of deaths equals the average number of births.

______ **3.** The average number of births is greater than the average number of deaths.

a. positive growth rate

b. negative growth rate

c. zero growth rate

4. Growth rate is the birth rate minus the

____.

5. Suppose that every year, one half of the population has two offspring per person, and the other half has none. If all members of the population die after a year, what is the resulting growth rate? Explain your answer.

____

____

____

SEQUENCING INFORMATION

One reading skill is the ability to sequence information, or to logically place items or events in the order in which they occur.

Sequence the statements below to illustrate zero population growth. Write "1" on the line in front of the first step, "2" on the line in front of the second step, and so on.

______ **6.** The population size returns to what it was in year x.

______ **7.** Two adults produce two offspring in year x.

______ **8.** The offspring, as adults, reproduce one offspring each.

______ **9.** The parents die.

RECOGNIZING SIMILARITIES AND DIFFERENCES

One reading skill is the ability to recognize similarities and differences between two phrases, ideas, or things. This is sometimes known as comparing and contrasting.

Read each question and write the answer in the space provided.

10. Explain the difference between negative growth rate and zero growth rate.

____

____

____

11. What is similar about negative growth rate and zero growth rate?

____

____

____

RECOGNIZING CAUSE AND EFFECT

One reading skill is the ability to recognize cause and effect.

Read the question and write the answer in the space provided.

12. What would be the result if a population did not replace its deaths with new births?

____

____

Name ______________________________ Class ______________ Date ____________

Skills Worksheet

Active Reading

Section: How Species Interact with Each Other

Read the passage below and answer the questions that follow.

An organism that lives in or on another organism and feeds on the other organism is a *parasite*. The organism the parasite takes its nourishment from is known as the *host*. The relationship between the parasite and its host is called **parasitism**. Examples of parasites are ticks, fleas, tapeworms, heartworms, bloodsucking leeches, and mistletoe.

Photos of parasites may make you feel uneasy, because parasites are somewhat like predators. The differences between a parasite and a predator are that a parasite spends some of its life in or on the host, and that a parasite does not usually kill its host. In fact, the parasite has an evolutionary advantage if it allows its host to live longer. However, the host is often weakened by or exposed to disease from the parasite.

IDENTIFYING MAIN IDEAS

One reading skill is the ability to identify the main idea of a passage. The main idea is the main focus or key idea. Frequently, a main idea is accompanied by supporting information that offers detailed facts about main ideas.

Read each question and write the answer in the space provided.

1. Give four examples of parasites.

____

____

2. What does a parasite get from its host?

____

3. What is the relationship between a parasite and its host called?

____

In the space provided, write the letter of the term or phrase that best completes each statement or best answers each question.

______ **4.** A parasite
- **a.** takes nourishment from another organism.
- **b.** always eventually kills its host.
- **c.** cannot live in mistletoe.
- **d.** All of the above

______ **5.** A host
- **a.** is like a predator.
- **b.** is the organism a parasite lives on or in.
- **c.** may make you feel uneasy.
- **d.** usually kills its parasite.

VOCABULARY DEVELOPMENT

Read each question and write the answer in the space provided.

6. The prefix *para-* means "alongside," while the Greek word *sitos* means "grain" or "food." Use this information to define *parasite.*

____

____

7. If the suffix *-ism* means "the practice of," how would you define *parasitism*?

____

____

RECOGNIZING SIMILARITIES AND DIFFERENCES

One reading skill is the ability to recognize similarities and differences between two phrases, ideas, or things. This is sometimes known as comparing and contrasting.

Read each question and write the answer in the space provided.

8. How are parasites and predators alike?

____

____

9. How are parasites and predators different?

____

____

RECOGNIZING CAUSE AND EFFECT

One reading skill is the ability to recognize cause and effect.

Read each question and write the answer in the space provided.

10. Why is it beneficial for a parasite to allow its host to live?

____

____

11. What effect does a parasite's presence usually have on its host?

____

____

Name ______________________ Class ______________ Date ____________

Skills Worksheet

Map Skills

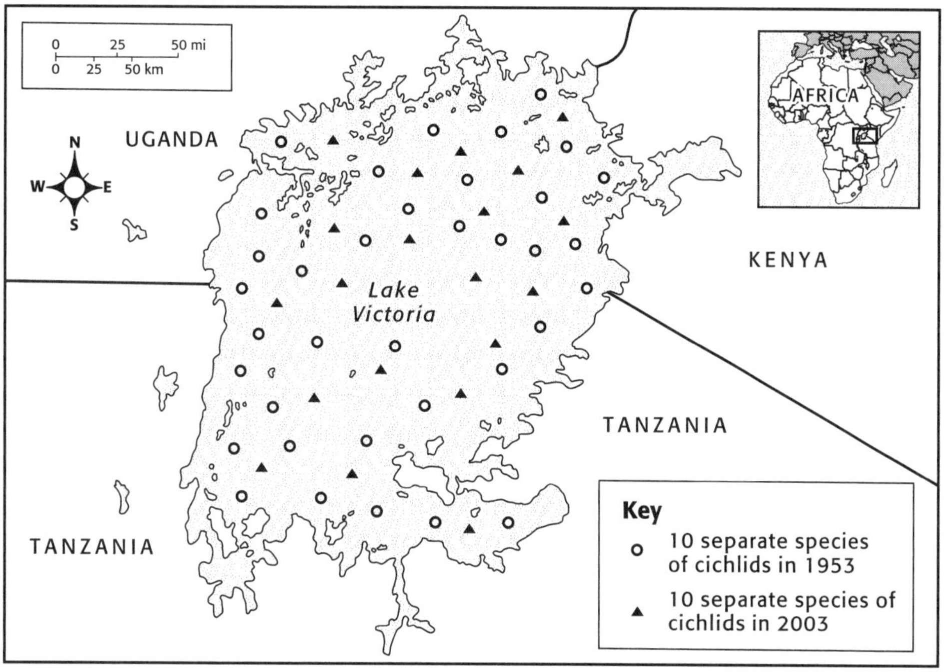

Lake Victoria, shown in the population map above, is the largest lake in Africa and is home to fish called cichlids (pronounced SIK-lids). In the 1950s, a new fish, the Nile Perch, was introduced into Lake Victoria. This map shows the change in the cichlid population since the Nile Perch was introduced.

Use the map above to answer the questions below.

1. **Using a Key** In which year was the number of cichlids the greatest?

____

2. **Using a Key** How many years did it take for the cichlid population to decrease by 50 percent?

____

3. **Inferring Relationships** Which fish in Lake Victoria do you think is higher on the food chain—the cichlid or the Nile Perch? Explain your answer.

____

4. **Identifying Trends** Based on the rate of decrease of cichlids shown on the map, what do you think will happen to the cichlid population by 2050? Explain your answer.

____

Name ______________________ Class ______________ Date ____________

Assessment

Quiz

Section: How Populations Change in Size

MATCHING

In the space provided, write the letter of the description that best matches the term or phrase.

______ **1.** number of individuals per unit area

______ **2.** example of indirect competition for resources

______ **3.** relative distribution of individuals

______ **4.** water, sunlight, or nutrients for plants

______ **5.** deaths caused by disease

a. density dependent
b. dispersion
c. density
d. territory
e. limiting resource

MULTIPLE CHOICE

In the space provided, write the letter of the term or phrase that best completes each statement or best answers each question.

______ **6.** Which of the following is one of the main properties used to describe a population?
a. number of individuals
b. color of individuals
c. number of species
d. kind of adaptations

______ **7.** For a population's growth rate to be zero
a. more deaths than births must occur.
b. more births than deaths must occur.
c. no births can occur.
d. the average number of births and deaths must be the same.

______ **8.** Exponential growth occurs when a population
a. exceeds the habitat's carrying capacity.
b. is competing with another species.
c. grows increasingly faster.
d. breeds with another species.

______ **9.** Which of the following limits a population's biotic potential?
a. a minimum number of offspring each pair can produce
b. a maximum number of offspring each individual can produce
c. the number of interactions each individual has
d. the size of offspring each individual can produce

______ **10.** Which of the following limits a population's growth?
a. carrying capacity of its habitat
b. natural disasters
c. severe weather
d. all of the above

Name ______________________ Class ______________ Date ____________

Assessment

Quiz

Section: How Species Interact with Each Other

MATCHING

Write the letter of the term or phrase that best matches the description.

______ **1.** Species A benefits and species B is killed.

______ **2.** Species A and B negatively affect each other.

______ **3.** Species A and B both benefit.

______ **4.** Species A benefits and species B is unaffected.

______ **5.** Species A benefits and species B is harmed but not killed.

a. competition
b. predation
c. parasitism
d. mutualism
e. commensalism

MULTIPLE CHOICE

In the space provided, write the letter of the term or phrase that best completes each statement or best answers each question.

______ **6.** Which of the following statements is correct?
a. An organism's niche is only the part of its habitat that it eats.
b. An organism's habitat is a location.
c. Habitat and niche are the same thing.
d. An organism's niche is outside its habitat.

______ **7.** Which of the following is part of an American bison's niche?
a. grasslands
b. gray wolf
c. water
d. all of the above

______ **8.** When two species in an area eat the same type of food but eat at different times of the day, their niches
a. are the same.
b. are examples of commensalism.
c. overlap.
d. eliminate competition.

______ **9.** If exponential growth occurs in the population of a species of predator, the population of its prey will most likely
a. decrease quickly.
b. increase slowly.
c. stay the same.
d. experience exponential growth.

______ **10.** If two species coevolve, they may develop adaptations that
a. reduce the harm of the relationship.
b. increase the benefit of the relationship.
c. prevent any relationships.
d. Both (a) and (b)

Name ______________________ Class ______________ Date ____________

Assessment

Chapter Test

Understanding Populations

MATCHING

Write the letter of the term or phrase that best matches the description.

______ **1.** kangaroo's role as a large herbivore on Australian grasslands		**a.** predation
______ **2.** woodpeckers eating at a birdfeeder		**b.** population
______ **3.** an owl snatching a mouse from a field to eat		**c.** clumped dispersion
______ **4.** cause of death that does not occur more quickly in crowded populations		**d.** competition
______ **5.** an orchid using a high tree branch as a place of attachment to receive more sunlight but not affecting the tree		**e.** generation time
______ **6.** all the black squirrels living in a forest		**f.** parasitism
______ **7.** three lampreys attached to a fish and sucking its body fluids for food		**g.** mutualism
______ **8.** average age at which members of a species reproduce		**h.** niche
______ **9.** description of a large population of geese gathered in a marsh		**i.** commensalism
______ **10.** a butterfly pollinating a flower as it drinks nectar from the flower		**j.** density independent

MULTIPLE CHOICE

In the space provided, write the letter of the term or phrase that best completes each statement or best answers each question.

______ **11.** The number of wild horses per square kilometer in a prairie is the horse population's

a. density.
b. dispersion.
c. size.
d. birth rate.

______ **12.** If over a long period of time each pair of adults in a population had only two offspring and the offspring lived to reproduce, the population would

a. grow.
b. shrink.
c. remain the same.
d. disperse randomly.

______ **13.** Which of the following species has the highest reproductive potential?
a. rabbit
b. elephant
c. human
d. horse

______ **14.** Which of the following is *not* an example of exponential growth?
a. rabbit populations after being introduced to Australia
b. reindeer of the Probilof Islands after eating most of the Lichens
c. a bank account that earns interest
d. mold appearing on bread overnight

______ **15.** The carrying capacity of an environment for a particular species at a particular time is determined by the
a. number of individuals in the species.
b. distribution of the population.
c. reproductive potential of the species.
d. supply of the most limited resources.

______ **16.** Competition for food *cannot* occur
a. between two populations.
b. among members of the same population.
c. among populations whose niches overlap.
d. between animals from two different ecosystems.

______ **17.** A bird that feeds at night and a bird that feeds during the day from the same flower is an example of
a. direct competition
b. mutualism
c. indirect competition
d. indirect commensalism

______ **18.** In which type of interaction between species does one species benefit by harming another species but not killing it?
a. predation
b. parasitism
c. mutualism
d. commensalism

______ **19.** Which of the following examples would be least likely to be considered a symbiotic interaction?
a. A wren builds a nest in a cactus.
b. A yucca moth pollinates and lays eggs on yucca flowers.
c. A kit fox hunts and feeds on a kangaroo rat.
d. Bacteria in a fox's digestive system help it digest food.

______ **20.** Which of the following two species represent a relationship that has coevolved?
a. flowering plants and their pollinators
b. foxes and their coyote competitors
c. house cats and their prey
d. rabbits and their relatives

Name ________________ Class ________ Date ________

Assessment

Chapter Test

Understanding Populations

MATCHING

In the space provided, write the letter of the term or phrase that best matches the description.

_____ **1.** organism's way of life

_____ **2.** arrangement of a population within a given space

_____ **3.** a group of individuals of the same species living in a particular place

_____ **4.** interaction in which one organism feeds upon another organism

_____ **5.** birth rate minus death rate

_____ **6.** organisms attempt to use same resources

_____ **7.** interaction in which one organism benefits and the other organism is unaffected

_____ **8.** maximum population an ecosystem can support indefinitely

_____ **9.** a way to reduce competition between species

_____ **10.** factor that determines the carrying capacity of an ecosystem

a. population
b. growth rate
c. competition
d. carrying capacity
e. limiting resource
f. dispersion
g. commensalism
h. niche restriction
i. niche
j. predation

MULTIPLE CHOICE

In the space provided, write the letter of the term or phrase that best completes each statement or best answers each question.

_____ **11.** Which of the following organisms has the highest reproductive potential?
a. dogs
b. elephants
c. bacteria
d. humans

_____ **12.** An example of a population would be all
a. trees in a forest.
b. red maple trees in a forest.
c. plants in a forest.
d. animals in a forest.

______ **13.** The density of a population is
a. the number of individuals born every year.
b. the proportion of males and females.
c. the number of individuals living in cities.
d. the number of individuals per unit area.

______ **14.** Each of the following is an example of a parasite *except*
a. a roundworm in a human's intestine.
b. a cow in a pasture.
c. a tick on a cat.
d. mistletoe on a tree.

______ **15.** The relationship between a Canadian lynx and a snowshoe hare is an example of
a. parasite and host.
b. predator and prey.
c. competition.
d. mutualism.

______ **16.** In which of the following types of interactions is neither species harmed?
a. predation
b. competition
c. parasitism
d. commensalism

______ **17.** Which of the following populations has a random dispersion?
a. flock of flamingos
b. pine trees in a pine forest
c. herd of bison
d. solitary snakes in a desert

______ **18.** Which of the following would be the *most* likely cause of a large number of density-independent deaths in a population?
a. winter storms
b. disease-carrying insects
c. predators
d. limited resources

______ **19.** Thick fur in deer is *not* an example of coevolution, because
a. thick fur is an adaptation.
b. deer with thick fur live longer.
c. thick fur evolved in response to a cold climate, not in response to other organisms.
d. in the lowlands, where the climate was sunny and warm, deer that did not have thick fur became separated from other deer that did have thick fur.

______ **20.** A species of plant has exponential growth after it is introduced into an area where it has never been. Which statement best describes exponential growth?
a. Each individual plant grows much larger than usual.
b. The population immediately decreases.
c. Within a few years the population increases dramatically.
d. The species' reproductive potential declines.

SHORT ANSWER

Write the answers to the following questions in the spaces provided.

21. The cardon and organ-pipe are flowering cacti that depend on bats for pollination. The bats pollinate the cacti as they eat the nectar in the cacti's flowers and spread its seeds when they eat the cactus fruit. Studies of the cacti show that they are not producing as much fruit as they could. It was also noted that bats living near these cacti had been driven from their cave homes by local villagers. What is the relationship between the bats and the cacti? How did the reduction in the number of bats affect the cacti?

____

____

____

____

22. If a population of rabbits experiences exponential growth, what might happen to the population of coyotes in the area? Explain your reasoning.

____

____

____

____

23. Predict what might happen to the population of rabbits and coyotes if the rabbits exceed the carrying capacity of the environment. Explain your reasoning.

____

____

____

24. Choose any two species with a close relationship that might have coevolved adaptations and describe how the adaptations are a benefit to both species.

____

____

____

____

____

ALTERNATIVE ASSESSMENT

25. The diagrams below show four different types of interactions between species. An arrow pointing from one organism to another means that the first organism has an effect on the second organism. Label each diagram with the correct type of interaction.

= Positive effect
= Negative effect
= No effect

Organism A 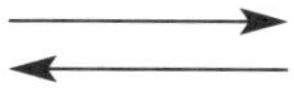 **Organism B**

a. ______________________________

Organism A 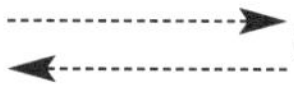 **Organism B**

b ______________________________

Organism A 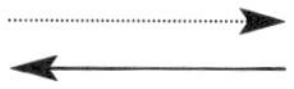 **Organism B**

c. ______________________________

Organism A 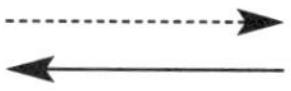 **Organism B**

d. ______________________________

or

____

Name ______________________ Class ______________ Date ____________

Skills Practice Lab

OBSERVATION

Studying Population Growth

You have learned that a population will keep growing until limiting factors slow or stop this growth. How do you know when a population has reached its carrying capacity? In this lab, you will observe the changes in a population of yeast cells. The cells will grow in a container and have limited food over several days.

OBJECTIVES

Observe, record, and **graph** the growth and decline of a population of yeast cells in an experimental environment.

Predict the carrying capacity of an environment for a population.

Infer the limiting resource of an environment.

MATERIALS

- compound microscope
- methylene blue solution, 1%
- micrometer, stage type or eyepiece disc for microscope
- microscope slide, with coverslip (5)
- pipet, 1 mL (5)
- test tube (5)
- yeast culture, in an Erlenmeyer flask (5)

Procedure

1. Your teacher will prepare several cultures of household baker's yeast (fungi of the genus *Saccharomyces*) in flasks. Each yeast culture will have grown for a different period of time in the same type of environment. Each flask will have been prepared with 500 mL of lukewarm water, 1 g of active dry yeast, and 20 g of sugar. The sugar is the only food source for the yeast.
2. Take a sample of yeast culture from the first flask (Time 0). Swirl the flask gently to mix the yeast cells evenly, and then immediately use the pipet to transfer 1 mL of yeast culture to a test tube. Add two drops of methylene blue solution to the test tube. The methylene blue will stain the dead yeast cells a deep blue but will not stain the living cells.
3. Make a wet mount by placing a small drop of your mixture of yeast and methylene blue on a microscope slide. Cover the slide with a coverslip.

4. Observe the mounted slide under the low power of a compound microscope. (Note: Adjust the light so that you can clearly see both stained and unstained cells.) After focusing, switch the microscope to high power (400 × or 1,000 ×).
5. Count the live (unstained) yeast cells and the dead (stained) cells that you see through the microscope. Use the micrometer ruler, or ask your teacher for the best counting method. Record the numbers of live cells in Table 1 and dead cells in Table 2.
6. Move the microscope slide slightly, and then make another count of the number of living and dead cells that you can see. Repeat this step until you have made four counts of the cells on the slide.
7. Calculate and record the average number of live cells per observation. Record this number in your data table. Do the same calculation for the dead cells.

TABLE 1: LIVE CELL COUNTS

	Cell Counts					
Time (h)	**1**	**2**	**3**	**4**	**Average**	**Class Average**
0						
12						
24						
36						
48						

TABLE 2: DEAD CELL COUNTS

	Cell Counts					
Time (h)	**1**	**2**	**3**	**4**	**Average**	**Class Average**
0						
12						
24						
36						
48						

8. Predict how many live and dead cells you expect to count in the samples from the other flasks. Record your prediction.
9. Repeat steps 3–8 for each of the flasks to obtain data that represents the growth of a population over a 48-hour period.

10. Clean up your work area, and store all lab equipment appropriately. Ask your teacher how to dispose of the yeast samples and any extra or spilled chemicals.

Analysis

1. **Analyzing Data** Share your data with the rest of the class. Calculate and record the class averages for each set of observations.

____

____

____

____

2. **Constructing Graphs** Graph the changes in the average numbers of live yeast cells and dead yeast cells over time. Plot the average number of cells per observation on the y-axis and the time (in hours from start of culture) on the x-axis.

3. **Describing Events** Describe the general population changes you observed in the yeast cultures over time.

____

____

____

____

Conclusions

4. **Evaluating Methods** Why were several counts taken and then averaged for each time period?

____

____

____

____

5. **Evaluating Results** Were your predictions of the yeast cell counts close to the actual average counts? How close were your predictions relative to the variation among all the samples?

____

____

____

____

6. **Applying Conclusions** Did the yeast cell populations appear to reach a certain carrying capacity? What was the limiting resource in the experimental environment of the flasks?

____

____

____

____

Extension

1. **Designing Experiments** Form a hypothesis about another factor that might limit the yeast's population growth, and explain how you would test this hypothesis.

____

____

____

____

____

Name ____________________ Class ____________ Date ____________

Skills Practice Lab

MODELING

Estimating Wild Animal Populations

One popular and simple technique for estimating a wild population of animals is called the mark-recapture method. It works like this: Suppose that you want to estimate the population of goldfish in a pond. You catch, tag, and release 40 fish. A few days later, you catch 40 fish and notice that 10 of the fish were tagged from the first catch—in other words, they were recaptured. To estimate the population of fish in the pond you can use a mathematical model. Multiply the number of fish in the first sample *(M)* by the number in the second sample *(n)*, and divide the product by the number of "recaptures" *(R)* to get the population of fish in the pond *(N)*:

$$\boxed{N = \frac{Mn}{R}} = \frac{\text{(first sample)} \times \text{(second sample)}}{\text{number recaptured}} = \text{estimated population}$$

To estimate the fish population,

$$\frac{40 \times 40}{10} = \frac{1600}{10} = 160$$

Therefore, the estimated number of goldfish in the pond is 160. For this model to give accurate estimates, you need to sample a fairly large population, and at least one animal must be captured in each sample. In general, the bigger your samples, the more accurate your estimate.

In this lab, you will practice the skill of estimating wild animal populations by setting up a model wild animal population in the lab. You will analyze the results of your lab and evaluate the use of the mark-recapture method as a good way to estimate a population.

OBJECTIVES

Construct a model of a wild animal population survey in the lab.

Survey and **estimate** wild animal populations using a pebble mark-recapture method laboratory model.

Analyze estimation data and evaluate the use of this method to estimate a population.

MATERIALS

- jar, 1-quart, plastic
- markers, 2 shades
- pebbles to fill jar

Procedure

PRACTICE USING THE MATHEMATICAL MODEL

1. You are an entomologist (a scientist who studies insects) trying to determine the population of Japanese beetles in your backyard. Two weeks ago you captured, marked, and released 100 beetles. Yesterday, you caught 40 beetles; 20 were recaptured from the first sample. Estimate the Japanese beetle population in your backyard. Show your work.

____

TRIAL 1—USING THE MARK-RECAPTURE METHOD

2. Fill a jar halfway with pebbles. These pebbles represent a population of wild animals. Do not count the pebbles.
3. Remove a handful of pebbles from the jar. The handful represents your first sample of animals. Count the pebbles, and write the total on the line below. Mark each pebble in your sample with one of the markers. Return the pebbles to the jar and thoroughly mix them with the others.

____

4. Remove another handful of pebbles from the jar, and record the total below.

____

5. Count and record the number of pebbles that were "recaptured."

____

TRIAL 2—USING THE MARK-RECAPTURE METHOD

6. Repeat steps 2–5 with the same jar of pebbles, but use a different marker. Record your data below.

   Number in first sample = ____________________

   Number in second sample = ____________________

   Number recaptured = ____________________

7. Count the total number of pebbles in the jar. Record the number below.

____

Analysis

1. **Organizing Data** Use Equation 1 to estimate the number of pebbles in the jar for both trial 1 and trial 2 data. Write your estimations on the line below.

____

2. Analyzing Data Compare the actual number of pebbles recorded in step 7 of the Procedure with the estimates above.

____

____

3. Analyzing Data Analyze the following data examples and determine which would reflect the largest population and which would reflect the smallest. Explain your answer.

a. large first sample, large second sample, large recapture

b. large first sample, large second sample, small recapture

c. small first sample, large second sample, large recapture

d. small first sample, small second sample, large recapture

____

____

____

____

____

____

4. Analyzing Data Analyze the following situation. You are surveying two ponds, one large and one small, for goldfish. You catch, tag, and release 20 goldfish from each pond. The next day, you catch 20 goldfish from each pond and count 8 recaptures from the small pond and 2 from the large pond.

a. Estimate the population of goldfish in the small pond.

____

b. Estimate the population of goldfish in the large pond.

____

c. Why would a large pond tend to have fewer recaptures than a small pond?

____

____

____

5. **Analyzing Data** Analyze the following situation. You captured, marked, and released 5 turtles from a pond, and caught 10 unmarked turtles the next day. Would you have enough information to estimate the population using the mark-recapture method?

____

____

____

6. **Analyzing Results** Distinguish the following situation from the previous situation. If you captured and marked one turtle from a pond and captured the same turtle the next day, can you conclude that only one turtle lives in the pond? Explain your answer.

____

____

____

____

Conclusions

7. **Drawing Conclusions** Draw your conclusion on the use of the mark-recapture method for estimating a population. Is it a good way of estimating populations? Explain your answer.

____

____

8. **Applying Conclusions** Imagine that you are studying birds that are flying south for the winter. How might their migration affect the results of a mark-recapture study? Can you accurately estimate the migrating bird population using the mark-recapture method? Explain your answer.

____

____

____

____

Name ______________________________ Class ______________ Date ____________

Skills Practice Lab

CBL™ PROBEWARE

Determining Growth Rate

Populations do not grow indefinitely. Their sizes fluctuate, and are determined by *emigration* (leaving a population), *immigration* (joining a new population), *natality* (births), and *mortality* (deaths). Changes in the environment and the availability of resources impact these factors.

In this lab, you will investigate a population of yeast. The population will be closed, and can change only by natality and mortality. You will use a colorimeter to determine how rapidly the yeast population grows. A colorimeter determines the concentration of a solution by measuring the amount of absorbed light from a beam shining through a sample. You will learn how to construct a standard curve that will allow you to determine the concentration of yeast cells/mL from the measured absorbance.

OBJECTIVES

Use a colorimeter to measure the light absorbance of a growth medium containing a population of yeast cells.

Graph a growth curve for a population of yeast cells.

Determine the concentration of yeast cells in your growth medium from a standard curve of absorbance versus yeast concentration.

Compare the yeast population growth curve to a logistic growth curve.

MATERIALS

- apple juice (growth medium)
- CBL System
- colorimeter cuvette with lid
- colorimeter
- coverslip
- culture tube, 20 × 150 mm and cap
- graduated cylinder, 10 mL
- graph paper
- lab apron
- link cable
- microscope
- microscope slide
- paper towel
- pipets, 15 cm disposable (2)
- safety goggles
- stopper, test-tube
- test tube, 15 mm × 125 mm (2)
- test-tube rack
- TI graphing calculator
- water, distilled
- yeast solution

Procedure

PREPARING THE GROWTH MEDIUM

1. Put on safety goggles and a lab apron.

2. Label a clean, dry 20 × 150 mm culture tube with your name and lab period. Add 30 mL of apple juice to the tube.

3. Obtain the yeast solution from your teacher. Swish the solution gently to mix it evenly. Then extract some into a pipet. Add 3 mL of the yeast solution to the apple juice in the tube. Pipet the medium gently up and down a few times to mix the juice and yeast solution. Leave your tube in a warm location, such as on a shelf.

SETTING UP THE CBL SYSTEM

4. Connect the CBL unit and the calculator with the link cable. Press the link cable firmly into each device to assure a good connection. Connect the colorimeter to Channel 1 of the CBL unit.

5. Turn on the CBL unit and the calculator. Start the CHEMBIO program. Go to the MAIN MENU.

6. Select SET UP PROBES. Enter "1" as the number of probes. Select COLORIMETER from the SELECT PROBE menu. Enter "1" as the channel number. You are now ready to set up the growth medium and calibrate the CBL unit and colorimeter.

CALIBRATING THE COLORIMETER

7. To avoid bubble formation on the cuvette sides, slowly fill a colorimeter cuvette $\frac{3}{4}$ full of distilled water. Wipe the outside dry with a paper towel. If bubbles are present, gently tap the cuvette with your finger to loosen them.

8. Place the cuvette into the colorimeter with the ribs facing the front and back of the colorimeter chamber, as shown in **Figure 1.** A smooth side of the cuvette should be facing the white reference mark on the colorimeter.

FIGURE 1 PLACE THE CUVETTE IN THE COLORIMETER

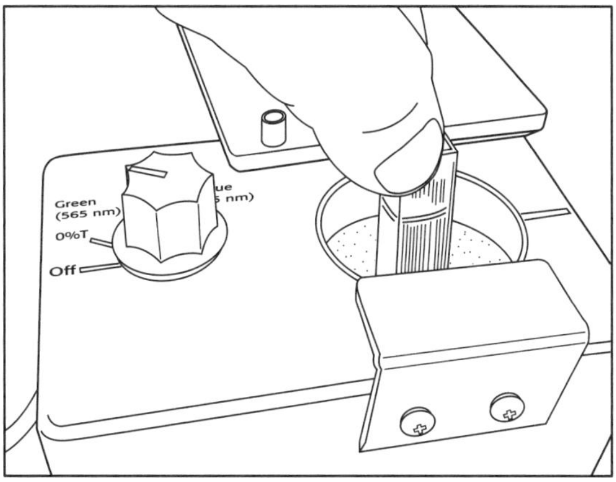

9. Close the colorimeter lid. Rotate the wavelength knob of the colorimeter to the "0 % T" position. When the display reading on the CBL unit is stable, press TRIGGER on the CBL unit.

10. On the graphing calculator, enter "0" as the reference. Turn the wavelength knob of the colorimeter to the Green LED position. When the CBL screen display is stable, press TRIGGER on the CBL unit. On the calculator, enter "100" as the reference. Leave the wavelength knob set to the Green LED setting to test samples.

11. On the calculator, press ENTER to return to the MAIN MENU. Select COLLECT DATA from the MAIN MENU. Select MONITOR INPUT from the DATA COLLECTION menu. On this setting, the CBL unit will monitor the light absorbance readings and display them on the graphing calculator.

MEASURING LIGHT ABSORBANCE

12. Agitate the culture tube gently to distribute the yeast organisms evenly. Remove the cap and withdraw about 2 mL of the growth medium with a sterile pipet. Fill a clean, dry cuvette with the growth medium from the pipet. Recap the culture tube immediately. Be sure the cuvette does not have any air bubbles in it. Wipe the outside dry with a paper towel. Rinse and dry the pipet.

13. Place the cuvette into the colorimeter. Close the lid and wait for the CBL display reading to stabilize. Read the absorbance value on the graphing calculator. If the absorbance reading is 1.0 or greater, follow the instructions in step 14 for diluting the growth medium sample. Record the absorbance reading and the dilution amount in **Table 1** for Day 1. If the growth medium was undiluted, record 1 for the dilution amount. If it was diluted by a factor of 10, record 0.1 for the dilution amount. If the first dilution was diluted, record 0.01 (1/100 dilution). Then, go to step 15.

14. Dilute your growth medium sample if the absorbance value on the colorimeter is greater than 1.0. Into a clean, dry test tube, place 1 mL of the growth medium from the cuvette. To the test tube containing the 1 mL of growth medium, add 9 mL of fresh juice. This produces a dilution of $\frac{1}{10}$ of the sample. Stopper the test tube, and shake gently to mix the contents. Remove 2 mL and place it into a clean cuvette. Wipe the cuvette and place it into the colorimeter. Follow the instructions in step 13 to measure the absorbance.

15. Turn off the CBL System by pressing "+" on the calculator, then QUIT. Remove the cuvette from the colorimeter, and save the solution to measure yeast concentration.

MEASURING THE YEAST CONCENTRATION

16. Place a lid on the cuvette, and shake it gently to mix the medium evenly. Using a clean pipet, remove some medium from the cuvette. Place two drops on a microscope slide. Cover the drops with a clean coverslip. Put the slide on a microscope. Focus on low power, change to high power, and refocus as necessary. Adjust the light to view the yeast cells most easily.

17. Count the number of yeast cells visible in the high power (HP) field of view. A yeast cell with a bud counts as two. Record the number of yeast cells in **Table 1** under "Yeast counted." If time permits or if your teacher directs you to do so, move the slide to observe another section of it, or make a second slide. Then make another HP field count of yeast cells. Average the two counts, and record the average in **Table 1** under "Yeast counted."

18. Clean and dry the microscope slide and coverslip. Empty the cuvette into the sink, then rinse, and dry the cuvette. Clean up your work area and wash your hands. Store your labeled culture tube in a test-tube rack in a warm location.

19. Repeat steps 4–18 each day for up to 8 days or until you notice the absorbance values have peaked and are in decline.

TABLE 1 DAILY ABSORBANCE AND YEAST CONCENTRATION MEASUREMENTS

Day	Measured absorbance (%)	Dilution	Absorbance (%)	Yeast counted (no. of cells)	Actual yeast (no. of cells)	Cell concentration (no./mL)
1						
2						
3						
4						
5						
6						
7						
8						
9						

Analysis

1. **Organizing Data** Calculate the absorbance and actual yeast population in a high power field for each day. Record your values in **Table 1.** Compute absorbance by dividing the measured absorbance in **Table 1** by the dilution. If the dilution is 1.0, then the absorbance value will equal the measured absorbance value.

   Example: If the measured absorbance is 0.121 and the dilution is 0.1,

   $$\frac{\text{measured absorbance}}{\text{dilution}} = \frac{0.121}{0.1} = 1.21$$

   Compute the actual yeast population by dividing the diluted yeast population by the dilution.

   Example: If the diluted yeast population is 90 and the dilution is 0.1,

   $$\text{actual yeast population} = \frac{\text{diluted yeast population}}{\text{dilution}} = \frac{90}{0.1} = 900$$

2. **Organizing Data** Calculate cell concentration, using the data from **Table 1.** The cell concentration can be determined as follows:

   $$\text{no. yeast cells/mL} = \frac{\text{(actual number of cells in one HP field of view)}}{0.000011 \text{ mL}}$$

   Record the values of cell concentration in **Table 1.** (0.000011 mL is the volume in one field of view)

3. **Constructing Graphs** Use the data in **Table 1** to construct two graphs. In **Figure 2,** graph absorbance (y-axis) against the number of days (x-axis). In **Figure 3,** graph yeast cell concentration (y-axis) against absorbance (x-axis). Only use data prior to the decline of the growth rate. Draw a line of best fit in **Figure 3.**

FIGURE 2 ABSORBANCE VERSUS TIME

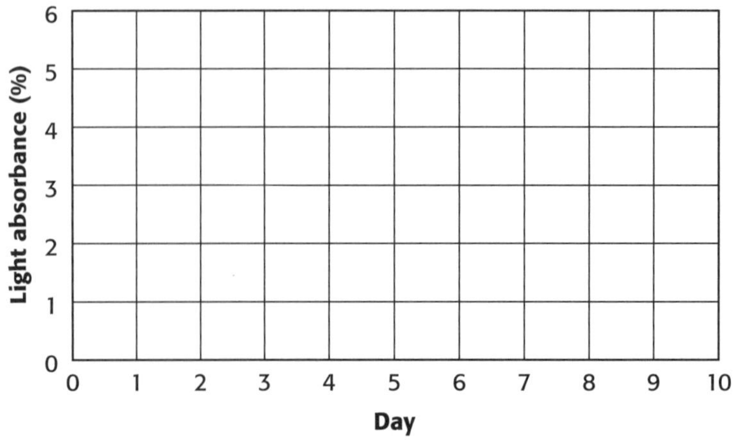

FIGURE 3 YEAST CONCENTRATION VERSUS ABSORBANCE

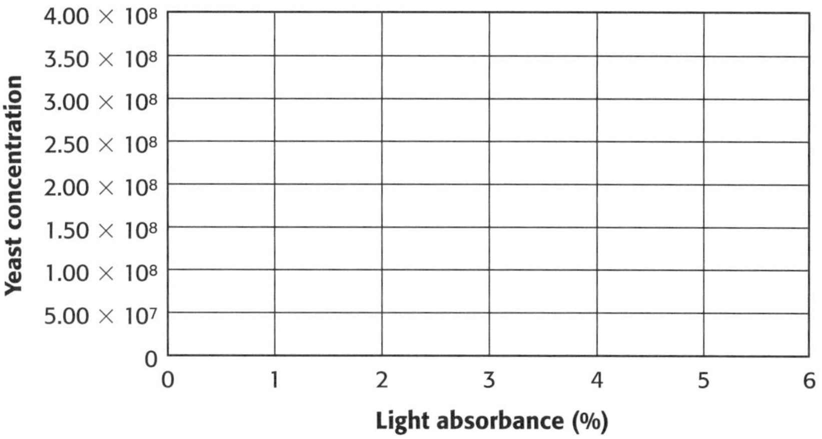

4. **Analyzing Graphs** Look at your graph of absorbance versus time. Describe the growth pattern of yeast during the experiment.

____

____

____

5. **Analyzing Graphs** The concentration of growing yeast is important to biologists because they often need to use the cells while they are still growing exponentially. Look at your graph of absorbance versus time. During what range of absorbance are these cells growing exponentially?

____

____

____

6. **Identifying Relationships** Biologists often use absorbance to determine the concentration of yeast and bacteria rather that doing actual cell counts each time. By using a standard curve such as the one you constructed in **Figure 3**, biologists can easily determine the concentration of microorganisms in growth media. What was the concentration of yeast in your sample when the absorbance was 1%? 2%?

____

____

____

Conclusions

1. **Analyzing Graphs** Suppose you want to collect the most yeast you can for an experiment in which the yeast need to be growing exponentially. At which absorbance would you collect your yeast?

____

____

____

2. **Evaluating Methods** During the dilution process, why was juice used rather than water?

____

____

3. **Evaluating Results** If there are no limits to the growth of a population, the population will grow exponentially. But because there are limited resources in nature, a population's growth will slow, and fluctuate around a number called its *carrying capacity*. The carrying capacity is the maximum population size an environment can support for a long period of time. This kind of growth is called logistic growth, and is modeled by the *logistic growth curve*, shown in **Figure 4.** Does your graph in **Figure 2** look like the logistic model? In what ways is it similar or different?

____

____

____

____

____

FIGURE 4 THE LOGISTIC POPULATION GROWTH MODEL

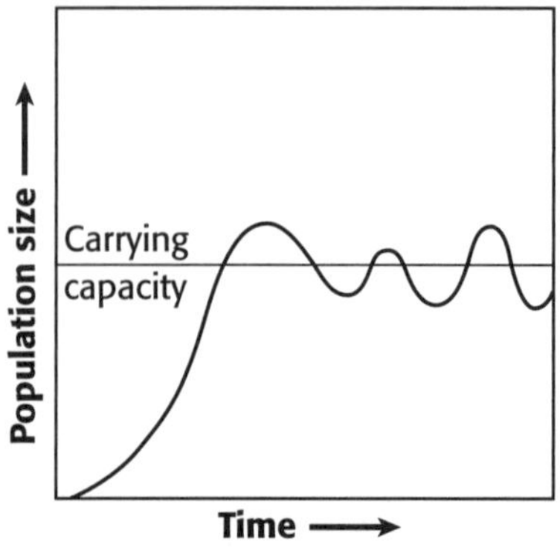

4. **Drawing Conclusions** Why do you think your graph in Figure 2 does not level off at, or fluctuate around, a carrying capacity? (Hint: Suppose that the graph in **Figure 4** represents the population growth of moose on an island. Consider the differences between the growth of the moose population on the island and the yeast population in the media when accounting for any differences in the graphs.)

____

____

____

____

____

5. **Drawing Conclusions** Describe the factors that contributed to the yeast population growth and decline.

____

____

____

____

____

Extensions

1. Design an experiment to test and investigate the effects of other growth mediums on yeast growth. Other clear juices, sugar solutions, or Sabouraud dextrose broth may be good choices for other growth mediums.
2. Use the library, media center, or Internet to research the growth of the human population. Find out when and why the human population began to grow rapidly. Learn about the past and current human population sizes and draw a graph of human population growth. Compare it to the logistic model. Develop a visual presentation for your class. Include information about population growth in developed and developing countries, and human carrying capacity.

Inquiry Lab

CONSUMER

Bug Off!

You are the lead research scientist for Bug Off! Corporation, the world's leading producer of insect repellents, and you have just received the following memo:

To: Director of Research and Development

From: Director of Marketing

cc: Director of Production

Subject: NEED FOR A NEW CONSUMER REPELLENT

We are getting reports from many states that the ant population is soaring and is causing many problems to homeowners. Existing ant repellents are not effective enough. We have an opportunity to increase our market share if we are first to the store shelves with a new product. Information I have says that Insects Be Gone!, our major competitor, is two weeks from production of their new product. Bug Off! needs to produce an effective, inexpensive and environmentally friendly ant repellent. During your research, please remember that every step of the scientific method must be considered and adapted to your development efforts.

I also urge you to use only common, readily accessible ingredients. This keeps the cost of materials down. The success of our efforts will also be determined by our decision to use nonhazardous, environmentally safe materials that help keep the cost of production low and protect the environment. Good luck!

OBJECTIVES

Collect data on the effects of your newly developed repellent on individual ants.

Construct and **test** a model of a home being invaded by ants to determine if the repellent is effective.

MATERIALS

- ants
- graph paper
- modeling supplies, common (cardboard, glue, adhesive tape, etc.)
- natural, nonhazardous ant repellent materials, such as peppermint, garlic, ground cinnamon, or spearmint plants
- standard lab equipment for measuring and mixing

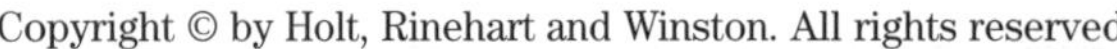

Procedure

1. Brainstorm with your lab team and develop a procedure that uses each of the steps of the scientific method and helps you meet your objectives. After you have designed the procedure, write the steps in the space provided below. The questions that start on the next page allow you to present key pieces of information that you should learn while following your procedure.

____

____

____

____

____

____

Analysis

1. **Constructing Graphs** Create a table, chart, or graph to document the data you collected on the effectiveness of the repellent on the ants. Present your table, chart, or graph on a separate sheet of graph paper.

2. **Analyzing Results** What percentage of individual ants were repelled by your product? Explain your analysis.

____

____

____

____

Conclusions

3. **Analyzing Graphs** Explain the data that shows the ability of your product to keep ants away. If the repellent worked on ants in the laboratory, will it work in a real life situation? Explain your analysis.

____

____

____

4. **Evaluating Models** Evaluate the effectiveness of your cardboard model. Use the space provided below to explain the model you used in pictures and words. Write the explanation on the lines provided below the space.

Extension

1. **Building Models** How effective would your repellent be if the ants were twice as big as those you used in your experiment and model?

Exploration Lab

LONG-TERM PROJECT

Calculating Generation Rate

Organisms compete for resources as they interact with their environment. Competition is a relationship in which different individuals or populations attempt to use the same limited resource. A resource is that which an organism needs to support itself to maintain its population. Resources include foods, open spaces, and hiding places. When organisms compete for resources, populations may be limited. Conditions are not resources, but are factors that may affect growth and reproduction.

Russian ecologist G. F. Gause conducted early experiments on competition. He grew *Paramecium aurelia* and *Paramecium caudatum* alone and together and provided them with food resources. Each species grew well alone, but when placed together, only *P. aurelia* survived. Gause then changed the conditions by adding sediment to the environment. The new conditions allowed *P. caudatum* to take refuge in the sediment. *P. caudatum* continued to multiply in this hiding place and coexisted with *P. aurelia*. Hiding in the sediment is an example of an adaptation to changing environmental conditions.

Results from many studies were summarized into what is known as the competitive exclusion principle. This principle states that two species cannot coexist indefinitely on the same limited resource. Organisms that are more adaptive tend to grow and displace those that do not adapt as well to changing conditions.

An interesting measure of growth is the generation rate. This is the number of generations of an organism that have been produced during a specified amount of time. The number of generations that have occurred, also called the generation number, can be calculated using the equation:

$$\text{Generation number (g)} = \frac{(\log B - \log A)}{\log 2}$$

"A" is a count of the cells at one time, and "B" is the count of the cells at a later time. The equation calculates the number of generations "g" that have occurred between the two times. Divide "g" by the elapsed time between the two counts, and you have determined the generation rate. Organisms with larger generation rates can overwhelm and outcompete other organisms.

In this lab, you will gather data and plot growth curves of noncompeting and competing populations of two types of organisms. You will analyze your data by comparing generation rates of the various populations.

OBJECTIVES

Experiment on organism competition.

Predict the effects of competition on number differences between competing populations.

Design an experiment to test the effects of competition on changing an environmental condition.

MATERIALS

- culture dish, six-well plastic with cover (3)
- distilled water, boiled and cooled (stock)
- microorganisms, *Blepharisma* (stock in glass bowl)
- microorganisms, *Euplotes* (stock in glass bowl)
- microscope, stereoscopic dissecting
- micropipettes, Pasteur, glass, fine-tipped, fitted with rubber bulbs
- paper, graph (2)
- water, bottled spring (stock)
- wheat germ kernel

Procedure

1. Label each well of a six-well plastic culture dish with the numbers 1 through 6. Fill each of the wells of the culture dish with 5 mL of bottled spring water. Add 0.5 mL of wheat germ kernel to each well.
2. Place one *Blepharisma* in each well. Place covers on each of the six wells and label the six-well plastic culture dish "*Blepharisma* Only."
3. Repeat step 1 using a separate six-well plastic culture dish. Place one *Euplotes* in each well. Place covers on each of the six wells and label this culture dish "*Euplotes* Only." Before beginning the experiment, make a hypothesis and record it below.

   ____

   ____

4. Count the number of cells per well daily for five days. Perform the count using a stereoscopic dissecting microscope. With some practice, you should be able to estimate the number of cells for each genus by scanning the contents of each well. Slow-moving ciliates such as *Blepharisma* are easy to count by eye. You can also use the micropipettes to do manual counts. Remove each organism as it is counted to another container while viewing through the stereoscopic dissecting microscope. When finished counting a well, be sure to return all organisms to the well from which you removed them.
5. Record the results of the counts of step 4 in Table 1.
6. Repeat step 1 using a separate six-well plastic culture dish. Place one individual cell of each of the two genera into each of wells 1 through 4.
7. Place two cells of the *Blepharisma* in well 5. This is a control population for the *Blepharisma*.

TABLE 1: *BLEPHARISMA* ENVIRONMENT POPULATION COUNTS

	1 day	2 days	3 days	4 days	5 days
Well 1					
Well 2					
Well 3					
Well 4					
Well 5					
Well 6					

TABLE 2: *EUPLOTES* ENVIRONMENT POPULATION COUNTS

	1 day	2 days	3 days	4 days	5 days
Well 1					
Well 2					
Well 3					
Well 4					
Well 5					
Well 6					

TABLE 3: MIXED/CONTROL ENVIRONMENT POPULATION COUNTS

	1 day	2 days	3 days	4 days	5 days
Well 1 ***Blepharisma***					
Well 1 ***Euplotes***					
Well 2 ***Blepharisma***					
Well 2 ***Euplotes***					
Well 3 ***Blepharisma***					
Well 3 ***Euplotes***					
Well 4 ***Blepharisma***					
Well 4 ***Euplotes***					
Well 5 ***Blepharisma***					
Well 5 ***Euplotes***					
Well 6 ***Blepharisma***					
Well 6 ***Euplotes***					

8. Place two cells of the *Euplotes* in well 6. This is the control population for the *Euplotes*.

9. Place covers on each of the six wells and label this culture dish "Mixed/Control."

10. Count the number of cells per well daily for five days. Follow the instructions of step 4 on how to perform a count. Record the results in Table 2.

Analysis

1. Constructing Graphs Use a separate sheet of graph paper to plot growth curves for the *Blepharisma* populations. Place the title "*Blepharisma* Populations" at the top of the graph paper. Using data in Table 1, plot the average number of *Blepharisma* you observe each day on the y-axis and the number of days on the x-axis. Label the curve "No competition." On this same graph, use the data in Table 3 to plot a curve of the average number of *Blepharisma* you observe each day in the mixed environments. Label this curve "Competition." On this same graph, use the data in Table 3 to plot a curve of the number of *Blepharisma* you observe each day in the control environment. Label this curve "Control." Use the data in Table 2 and Table 3 and another piece of graph paper to plot growth curves for *Euplotes* populations. All averages are to be calculated by adding the counts in each well for a given day and dividing by the number of wells.

2. Organizing Data Calculate a generation rate for each of the six curves on your graphs. Show all calculations.

____

____

____

____

____

____

Conclusions

3. Drawing Conclusions Draw conclusions as to which organism will outcompete the other in the mixed environment. What do the generation rates for each of the six population curves suggest about this competition? How does this compare to your hypothesis?

____

____

____

____

____

____

____

4. Making Predictions Imagine that you created a mixed environment starting with five times more of the organism with the lower generation rate. Do you think this would change the results of a competition? Explain.

____

____

____

____

____

____

Extension

1. Designing Experiments Design an experiment (objective, materials, and procedure) that tests the effect of a change in environmental conditions on the two microorganisms. Decide which condition you will change compared to the experiment you already performed. You could use G. F. Gause's experiment as an example. Make sure that you include unmixed, mixed, and control populations in your experiment.

Lesson Plan

Section: How Populations Change in Size

Pacing

1 block = 45 minutes

Regular Schedule	**with lab(s):** 3 days	**without lab(s):** 2 days
Block Schedule	**with lab(s):** 1.5 days	**without lab(s):** 1 day

Objectives

1. Describe the three main properties of a population.
2. Describe exponential population growth.
3. Describe how the reproductive behavior of individuals can affect the growth rate of their population.
4. Explain how population sizes in nature are regulated.

National Science Education Standards Covered

UCP 1: Systems, order, and organization.

UCP 3: Change, constancy, and measurement.

UCP 4: Evolution and equilibrium.

UCP 5: Form and function.

LS 4c: Organisms both cooperate and compete in ecosystems. The interrelationships and interdependencies of these organisms may generate ecosystems that are stable for hundreds or thousands of years.

LS 4d: Living organisms have the capacity to produce populations of infinite size, but environments and resources are finite. This fundamental tension has profound effects on the interactions between organisms.

SPSP 2a: Populations grow or decline through the combined effects of births and deaths, and through emigration and immigration. Populations can increase through linear or exponential growth, with effects on resource use and environmental pollution.

SPSP 2c: Populations can reach limits to growth. Carrying capacity is the maximum number of individuals that can be supported in a given environment. The limitation is not the availability of space, but the number of people in relation to resources and the capacity of earth systems to support human beings. Changes in technology can cause significant changes, either positive or negative, in carrying capacity.

KEY
SE = Student Edition
ATE = Annotated Teacher Edition
CRF = Chapter Resource file

Block 1

FOCUS *5 minutes*

- ❑ **Using The Figure** ATE. Students examine the photo in the chapter opener and discuss the relationship between orcas and sea lions.
- ❑ **Bellringer** Bellringer Transparency, ATE. In their *EcoLog*, have students write a definition of a population and list examples of populations in their neighborhood.

MOTIVATE *5 minutes*

- ❑ **Activity** Two Types of Growth, ATE. Ask students to choose between two payment plans given in the ATE. Relate the plans to types of population growth. **(GENERAL)**

TEACH *35 minutes*

- ❑ **Reading Warm-Up** SE. Students write answers to questions 1–2 in their *EcoLog*.
- ❑ **Teaching Transparency** Population Change and Exponential Growth. Students use the illustration to see what happens to populations that grow exponentially.
- ❑ **QuickLab** Population Growth, SE. Students model and graph the population change in a species over a period of ten years. **(GENERAL)**
- ❑ **Using the Figure** Exponential Growth, ATE. Students use Figure 4 to learn about exponential growth and to compare it to linear growth. **(GENERAL)**

HOMEWORK *30 minutes*

- ❑ **Homework** Favorite Populations, ATE. Students find out about the population of their favorite plant or animal and record their findings in their *EcoLog*.

Block 2

LAB *45 minutes*

- ❑ **Modeling Lab** Estimating Wild Animal Populations, CRF. Students use mathematical formulas to estimate the wild population of animals. **(GENERAL)**

Block 3

TEACH *35 minutes*

- ❑ **Teaching Transparency** Population Changes and Carrying Capacity. Students use this graphic to study how populations increase and decrease around the carrying capacity of an ecosystem.
- ❑ **Using the Figure** What Limits Population Growth? ATE. Students study Figure 5 and discuss how the environment affects population growth. **(BASIC)**
- ❑ **Internet Activity** Island Carrying Capacities, ATE. Students research a population on an island and write an essay about their findings. **(GENERAL)**

- ❑ **Math Practice** Growth Rate, SE. Students calculate the growth rate of a population over time. (General)

CLOSE *10 minutes*

- ❑ **Section Quiz** Section 1, CRF. Students answer 10 questions that review the lesson content. **(General) Also in Spanish**
- ❑ **Concept Review Worksheet** How Populations Change in Size, CRF, Study Guide. Complete worksheet for this section. **(General) Also in Spanish**

HOMEWORK *30 minutes*

- ❑ **Section Review** How Populations Change in Size, SE. Assign questions 1–6 for review, homework, or as a quiz.
- ❑ **Active Reading Worksheet** Section 1, CRF. This exercise assesses reading comprehension of the material covered in the section. **(Basic)**

OTHER RESOURCE OPTIONS

- ❑ **Inclusion Strategies** ATE. Students model population growth and available resources using index cards.
- ❑ **Skills Practice Lab: Observation** Studying Population Growth, SE. Students observe the changes in a population of yeast cells and note whether a carrying capacity is reached. **(General)**
- ❑ **Datasheets for In-Text Lab:** Studying Population Growth, CRF. Students complete datasheets for In-Text Skills Practice Lab. **(General)**
- ❑ **CNN Presents Science in the News:** Science, Technology & Society, Segment 25, “Salmon Sound Barrier.” This segment focuses on the salmon population.
- ❑ **CNN Presents Science in the News:** Science, Technology & Society, Critical Thinking Worksheet, Segment 25, “Salmon Sound Barrier.” This worksheet reinforces the content of the corresponding CNN video.
- ❑ **go.hrw.com** For worksheets, videos, and other teaching aids related to this chapter, visit the HRW Web site and type in the keyword **HE4 POC.**
- ❑ **Video Select** Videos related to the chapter topics may be found at **go.hrw.com.** Type in the keyword **HE4 POCV.**
- ❑ **Internet Connect** Populations and Communities, SciLinks code: HE4085. Students use Internet sources to learn about populations in communities.
- ❑ **Guided Reading Audio CD Program** Understanding Populations Script. Assign Section 1. The audio program is a reading of the chapter content for ELL students, auditory learners, and struggling readers.

Lesson Plan

Section: How Species Interact with Each Other

Pacing

1 block = 45 minutes

Regular Schedule	**with lab(s):** N/A	**without lab(s):** 2 days
Block Schedule	**with lab(s):** N/A	**without lab(s):** 1 day

Objectives

1. Explain the difference between niche and habitat.
2. Give examples of parts of a niche.
3. Describe the five major types of interactions between species.
4. Explain the difference between parasitism and predation.
5. Explain how symbiotic relationships may evolve.

National Science Education Standards Covered

UCP 1: Systems, order, and organization.

UCP 4: Evolution and equilibrium.

UCP 5: Form and function.

LS 4c: Organisms both cooperate and compete in ecosystems. The interrelationships and interdependencies of these organisms may generate ecosystems that are stable for hundreds or thousands of years.

LS 4d: Living organisms have the capacity to produce populations of infinite size, but environments and resources are finite. This fundamental tension has profound effects on the interactions between organisms.

KEY
SE = Student Edition
ATE = Annotated Teacher Edition
CRF = Chapter Resource file

Block 4

FOCUS *5 minutes*

❑ **Bellringer** Bellringer Transparency, ATE. Students write a paragraph about a carnivore through the eyes of its prey.

MOTIVATE *5 minutes*

❑ **Activity** Constructing a Personal Niche Map, ATE. Ask students to think about how and where they fit in their community. Have them construct a personal niche map. **(BASIC)**

TEACH *35 minutes*

- ❑ **Teaching Transparency** Types of Species Interactions. Students use the graphic to learn the five types of species interaction.
- ❑ **Using the Figure** Indirect Interactions, ATE. Using Figure 10, students should note that four of the five interactions are shown in the picture. They brainstorm other examples of how the organisms might interact. **(GENERAL)**
- ❑ **Group Activity** Species Interactions Skit, ATE. Each of five groups of students plans a skit to illustrate one type of species interaction. The other groups try to guess which interaction the skit shows. **(GENERAL)**
- ❑ **Teaching Transparency** Niche Restriction Due to Competition. Students study the graphic to learn how competition causes the partitioning of potential niches.
- ❑ **Internet Activity** Restricted and Potential Niches, ATE. Students research online Robert MacArthur's study of niches, and illustrate his findings. **(GENERAL)**

HOMEWORK *20 minutes*

- ❑ **Homework** Artistic Interactions, ATE. Have students create a piece of art depicting one of the five types of species relationships. **(GENERAL)**
- ❑ **Map Skills Worksheet** Tracking Cichlids, CRF. Students use the map to speculate on the demise of cichlid species in Lake Victoria. **(GENERAL)**

Block 5

TEACH *35 minutes*

- ❑ **Using the Figure** Specialists Vs. Generalists, ATE. Discuss predators that are specialists and generalists. Use Figure 13 to show how the populations of specialized predators are tied to that of their prey. **(GENERAL)**
- ❑ **Case Study** Predator-Prey Adaptations, SE. Students read the case study and answer the two Critical Thinking questions in their *EcoLog*. **(GENERAL)**
- ❑ **Skill Builder** Graphing, ATE. Students simulate how prey population density affects the predator's success. Record the results in a table and on a graph. **(BASIC)**
- ❑ **Points of View** Where Should the Wolves Roam? SE. Students read about two perspectives on reintroducing wolves in certain places, then discuss their own opinions and answer the "What Do You Think?" questions. **(GENERAL)**

CLOSE *10 minutes*

- ❑ **Section Quiz** Section 2, CRF. Students answer 10 questions that review the lesson content. **(GENERAL) Also in Spanish**
- ❑ **Concept Review Worksheet** How Species Interact with Each Other, CRF, Study Guide. Complete worksheet for this section. **(GENERAL) Also in Spanish**

HOMEWORK *30 minutes*

- ❑ **Section Review** How Species Interact With Each Other, SE. Assign questions 1–5 for review, homework, or as a quiz.
- ❑ **Active Reading Worksheet** Section 2, CRF. This exercise assesses reading comprehension of the material covered in the section. **(BASIC)**

OTHER RESOURCE OPTIONS

- ❑ **Inclusion Strategies** ATE. Students make a poster chart about predation.
- ❑ **Field Activity** Observing Competition, SE. Students make a bird feeder and observe the birds that visit for several days. They record what they see in their *EcoLog*. **(GENERAL)**
- ❑ **Reading Skill Builder** Brainstorming, ATE. Have students copy the table from Figure 10 into their *EcoLog* and add examples of each type of interaction. **(BASIC)**
- ❑ **Homework** Predatory Strategies, ATE. Have pairs of students research a predator-prey relationship and present their findings to the class. **(GENERAL)**
- ❑ **Alternate Assessment** Parasitic Lifestyles, ATE. Students create a poster or report on a parasitic interaction. **(GENERAL)**
- ❑ **CBL™ Probeware Lab** Carrying Capacity, CRF. Students experiment with carrying capacity. **(ADVANCED)**
- ❑ **Consumer Lab** Bug Off! CRF. Groups of students create and test a bug spray using non-toxic common household products. **(GENERAL)**
- ❑ **Long-Term Project** Calculating Generation Rate, CRF. Students gather data and plot growth curves of non-competing and competing populations of two types of organisms. Students analyze data by comparing generation rates of the various populations. **(GENERAL)**
- ❑ **go.hrw.com** For worksheets, videos, and other teaching aids related to this chapter, visit the HRW Web site and type in the keyword **HE4 POC.**
- ❑ **Video Select** Videos related to the chapter topics may be found at **go.hrw.com.** Type in the keyword **HE4 POCV.**
- ❑ **Internet Connect** Coevolution, SciLinks code: HE4014. Students use Internet sources to learn about how organisms in close associations may coevolve.
- ❑ **Guided Reading Audio CD Program** Understanding Populations Script. Assign Section 2. The audio program is a reading of the chapter content for ELL students, auditory learners, and struggling readers.

Lesson Plan

Review and Assessment

Pacing

1 block = 45 minutes

Regular Schedule	**with lab(s):** N/A	**without lab(s):** 2 days
Block Schedule	**with lab(s):** N/A	**without lab(s):** 1 day

Block 6

REVIEW

- ❑ **Chapter Review** Understanding Populations, SE. Assign questions 1–34 to review the material for this chapter. Use the assignment guide to customize review for lessons covered.
- ❑ **Concept Review Worksheet** Understanding Populations, CRF, Study Guide. Complete worksheet for this chapter. **(GENERAL) Also in Spanish**
- ❑ **Critical Thinking Worksheets** Understanding Populations, CRF. Assign these worksheets to challenge students to think critically about the material in this chapter. **(ADVANCED)**

Block 7

ASSESSMENT

- ❑ **Chapter Test** Understanding Populations, CRF. Assign questions for general level chapter assessment. **(GENERAL) Also in Spanish**
- ❑ **Chapter Test** Understanding Populations, CRF. Assign questions for advanced level chapter assessment. **(ADVANCED)**

ALTERNATIVE ASSESSMENT

- ❑ **Alternative Assessment** Species Management Plan, ATE. Assign this activity for general level chapter assessment. **(GENERAL)**
- ❑ **Alternative Assessment** Species Interactions, ATE. Assign this activity for general level chapter assessment. **(GENERAL)**
- ❑ **Alternative Assessment** Competition Debate, ATE. Assign this activity for general level chapter assessment. **(GENERAL)**
- ❑ **Test Generator** One-Stop Planner. Create a customized homework assignment, quiz, or test using the HRW Test Generator program.
- ❑ **Test Item Listing** Understanding Populations, CRF. Use the Test Item Listing to identify questions to use in a customized homework assignment, quiz, or test.

Skills Practice Lab **OBSERVATION**

Studying Population Growth

Classroom Tested & Approved

Linda Culp
Thorndale
High School
Thorndale, Texas

Teacher Notes

TIME REQUIRED Two 45-minute class periods (48 hours)

SKILLS ACQUIRED

Collecting data
Communicating
Designing experiments
Experimenting
Inferring
Measuring
Organizing and analyzing data
Predicting

RATING Easy ← 1 2 3 4 → Hard

Teacher Prep–3
Student Set-Up–2
Concept Level–1
Clean Up–2

THE SCIENTIFIC METHOD

Make Observations Procedure, steps 5–10

Draw Conclusions Conclusions, questions 4–6

MATERIALS

The materials listed on the student page are enough for a group of 3 to 4 students. A stage micrometer, available from the laboratory supplies is a microscopic scale printed on a microscope slide. Similar micrometer discs that fit into the microscope eyepiece are also available.

To prepare the yeast cultures, obtain 3 standard supermarket packages (7 g) of active dry yeast (not fast-rising or instant yeast). Activate each of 5 batches of yeast culture at 12-hour intervals prior to the time students will start their observations (0, 12, 24, 36, and 48 hours). To activate the yeast, dissolve 20 g of sugar in 500 mL of warm water (approximately 32°C or 90°F) in a flask, then add 1 g (~1/4 tsp) yeast and swirl thoroughly. Keep the flasks covered in a warm, dark area until beginning the lab.

SAFETY CAUTIONS

Caution students to treat all microorganisms as potential pathogens, and tell them that methylene blue is harmful if swallowed and can cause skin or eye irritation. Call the Poison Control Center if a student ingests any of this substance. Remind students to wear gloves and goggles, to keep their hands away from their faces and to wash their hands after this lab. Review sterile techniques.

DISPOSAL

The yeast/methylene blue mixture should be disposed of as hazardous waste. Microscope slides and cover slips should be disposed of as contaminated glassware.

TECHNIQUES TO DEMONSTRATE

If the students have never used a microscope, you will have to demonstrate the use of the microscope to the students. You may also have to demonstrate techniques for counting microorganisms under the microscope.

TIPS AND TRICKS

Tell students that they are observing yeast at several stages of growth, to simulate the observation of a single population over time. Point out the generation of gas bubbles by the actively-growing cultures. Demonstrate for students how a culture is prepared (Time 0), how to properly swirl the flask of yeast (failure to evenly distribute cells will alter counts), how to use a pipet to transfer a sample into a test tube, how to place a small drop of yeast-stain mixture into a slide, and how to add the cover slip, and how to focus the microscope.

Tell students that the methylene blue only stains the dead yeast cells because living cells actively keep out the stain. Tell the students to keep the buds on actively-dividing yeast cells in their cell counts. You may want to experiment with and demonstrate a method of counting the cells that works best for the microscopes and materials you have. Variation among counts taken from the same flask should be less significant than differences between flasks at different stages.

Name ________________________ Class ______________ Date ____________

Skills Practice Lab

OBSERVATION

Studying Population Growth

You have learned that a population will keep growing until limiting factors slow or stop this growth. How do you know when a population has reached its carrying capacity? In this lab, you will observe the changes in a population of yeast cells. The cells will grow in a container and have limited food over several days.

OBJECTIVES

Observe, record, and **graph** the growth and decline of a population of yeast cells in an experimental environment.

Predict the carrying capacity of an environment for a population.

Infer the limiting resource of an environment.

MATERIALS

- compound microscope
- methylene blue solution, 1%
- micrometer, stage type or eyepiece disc for microscope
- microscope slide, with coverslip (5)
- pipet, 1 mL (5)
- test tube (5)
- yeast culture, in an Erlenmeyer flask (5)

Procedure

1. Your teacher will prepare several cultures of household baker's yeast (fungi of the genus *Saccharomyces*) in flasks. Each yeast culture will have grown for a different period of time in the same type of environment. Each flask will have been prepared with 500 mL of lukewarm water, 1 g of active dry yeast, and 20 g of sugar. The sugar is the only food source for the yeast.
2. Take a sample of yeast culture from the first flask (Time 0). Swirl the flask gently to mix the yeast cells evenly, and then immediately use the pipet to transfer 1 mL of yeast culture to a test tube. Add two drops of methylene blue solution to the test tube. The methylene blue will stain the dead yeast cells a deep blue but will not stain the living cells.
3. Make a wet mount by placing a small drop of your mixture of yeast and methylene blue on a microscope slide. Cover the slide with a coverslip.

Name ______________________________ Class ______________ Date ____________

Studying Population Growth *continued*

4. Observe the mounted slide under the low power of a compound microscope. (Note: Adjust the light so that you can clearly see both stained and unstained cells.) After focusing, switch the microscope to high power (400 × or 1,000 ×).

5. Count the live (unstained) yeast cells and the dead (stained) cells that you see through the microscope. Use the micrometer ruler, or ask your teacher for the best counting method. Record the numbers of live cells in Table 1 and dead cells in Table 2.

6. Move the microscope slide slightly, and then make another count of the number of living and dead cells that you can see. Repeat this step until you have made four counts of the cells on the slide.

7. Calculate and record the average number of live cells per observation. Record this number in your data table. Do the same calculation for the dead cells.

TABLE 1: LIVE CELL COUNTS

	Cell Counts					
Time (h)	**1**	**2**	**3**	**4**	**Average**	**Class Average**
0						
12						
24						
36						
48						

TABLE 2: DEAD CELL COUNTS

	Cell Counts					
Time (h)	**1**	**2**	**3**	**4**	**Average**	**Class Average**
0						
12						
24						
36						
48						

8. Predict how many live and dead cells you expect to count in the samples from the other flasks. Record your prediction.

9. Repeat steps 3–8 for each of the flasks to obtain data that represents the growth of a population over a 48-hour period.

10. Clean up your work area, and store all lab equipment appropriately. Ask your teacher how to dispose of the yeast samples and any extra or spilled chemicals.

Analysis

1. **Analyzing Data** Share your data with the rest of the class. Calculate and record the class averages for each set of observations.

   Answers may vary. Student counts should be similar to the overall average. Failure to stir the yeast culture may alter counts.

2. **Constructing Graphs** Graph the changes in the average numbers of live yeast cells and dead yeast cells over time. Plot the average number of cells per observation on the y-axis and the time (in hours from start of culture) on the x-axis.

   Students should create a double line graph to compare how the density of live and dead cells changes over time.

3. **Describing Events** Describe the general population changes you observed in the yeast cultures over time.

   Answers may vary. Yeast populations probably rose exponentially, reached a plateau, and then dropped quickly.

Conclusions

4. **Evaluating Methods** Why were several counts taken and then averaged for each time period?

   Counting several random samples helps to control for variation in yeast dispersion within the flask or within the microscope slide.

5. **Evaluating Results** Were your predictions of the yeast cell counts close to the actual average counts? How close were your predictions relative to the variation among all the samples?

   Answers may vary.

6. **Applying Conclusions** Did the yeast cell populations appear to reach a certain carrying capacity? What was the limiting resource in the experimental environment of the flasks?

   Answers may vary. If the yeast reached a maximum population size, this would indicate the carrying capacity. The limiting resources are food (sugar) and uncontaminated habitat.

Extension

1. **Designing Experiments** Form a hypothesis about another factor that might limit the yeast's population growth, and explain how you would test this hypothesis.

   Answers may vary.

Skills Practice Lab

MODELING

Estimating Wild Animal Populations

Teacher Notes

CLASSROOM TESTED & APPROVED

Alonda Droege
Evergreen
High School
Seattle, Washington

TIME REQUIRED One 45-minute class period

SKILLS ACQUIRED

Collecting data
Constructing models
Experimenting
Organizing and analyzing data

RATING Easy ← 1 2 3 4 → Hard

Teacher Prep–1
Student Set-Up–1
Concept Level–2
Clean Up–1

THE SCIENTIFIC METHOD

Make Observations Procedure, steps 2–7

Analyze the Results Analysis, questions 1–6

Draw Conclusions Conclusions, questions 7 and 8

MATERIALS

The materials for this lab are enough for a group of three students and easy to obtain. They can buy medium-size pebbles from a local gardening, home supply, or pet store sufficient to fill a 1-quart jar.

Name ______________________ Class ______________ Date ____________

Skills Practice Lab

MODELING

Estimating Wild Animal Populations

One popular and simple technique for estimating a wild population of animals is called the mark-recapture method. It works like this: Suppose that you want to estimate the population of goldfish in a pond. You catch, tag, and release 40 fish. A few days later, you catch 40 fish and notice that 10 of the fish were tagged from the first catch—in other words, they were recaptured. To estimate the population of fish in the pond you can use a mathematical model. Multiply the number of fish in the first sample *(M)* by the number in the second sample *(n)*, and divide the product by the number of "recaptures" *(R)* to get the population of fish in the pond *(N)*:

$$\boxed{N = \frac{Mn}{R}} = \frac{\text{(first sample)} \times \text{(second sample)}}{\text{number recaptured}} = \text{estimated population}$$

To estimate the fish population,

$$\frac{40 \times 40}{10} = \frac{1600}{10} = 160$$

Therefore, the estimated number of goldfish in the pond is 160. For this model to give accurate estimates, you need to sample a fairly large population, and at least one animal must be captured in each sample. In general, the bigger your samples, the more accurate your estimate.

In this lab, you will practice the skill of estimating wild animal populations by setting up a model wild animal population in the lab. You will analyze the results of your lab and evaluate the use of the mark-recapture method as a good way to estimate a population.

OBJECTIVES

Construct a model of a wild animal population survey in the lab.

Survey and **estimate** wild animal populations using a pebble mark-recapture method laboratory model.

Analyze estimation data and evaluate the use of this method to estimate a population.

MATERIALS

- jar, 1-quart, plastic
- markers, 2 shades
- pebbles to fill jar

Name ______________________ Class ______________ Date ____________

Procedure

PRACTICE USING THE MATHEMATICAL MODEL

1. You are an entomologist (a scientist who studies insects) trying to determine the population of Japanese beetles in your backyard. Two weeks ago you captured, marked, and released 100 beetles. Yesterday, you caught 40 beetles; 20 were recaptured from the first sample. Estimate the Japanese beetle population in your backyard. Show your work.

   $N = 100 \times \frac{40}{20} = 200$; the estimated population is 200 beetles.

TRIAL 1—USING THE MARK-RECAPTURE METHOD

2. Fill a jar halfway with pebbles. These pebbles represent a population of wild animals. Do not count the pebbles.
3. Remove a handful of pebbles from the jar. The handful represents your first sample of animals. Count the pebbles, and write the total on the line below. Mark each pebble in your sample with one of the markers. Return the pebbles to the jar and thoroughly mix them with the others.

   Answers may vary.
4. Remove another handful of pebbles from the jar, and record the total below.

   Answers may vary.
5. Count and record the number of pebbles that were "recaptured."

   Answers may vary.

TRIAL 2—USING THE MARK-RECAPTURE METHOD

6. Repeat steps 2–5 with the same jar of pebbles, but use a different marker. Record your data below.

   Number in first sample = **Answers may vary.**

   Number in second sample = **Answers may vary.**

   Number recaptured = **Answers may vary.**
7. Count the total number of pebbles in the jar. Record the number below.

   Answers may vary.

Analysis

1. **Organizing Data** Use Equation 1 to estimate the number of pebbles in the jar for both trial 1 and trial 2 data. Write your estimations on the line below.

   Answers may vary. Sample answer: trial 1 = 52, trial 2 = 49

Name ______________________ Class ______________ Date ____________

Estimating Wild Animal Populations *continued*

2. **Analyzing Data** Compare the actual number of pebbles recorded in step 7 of the Procedure with the estimates above.

Answers may vary. The actual number will probably be close but not identical to the estimated numbers.

3. **Analyzing Data** Analyze the following data examples and determine which would reflect the largest population and which would reflect the smallest. Explain your answer.

   a. large first sample, large second sample, large recapture

   b. large first sample, large second sample, small recapture

   c. small first sample, large second sample, large recapture

   d. small first sample, small second sample, large recapture

Answers may vary. Sample answer: b. reflects the largest population because there were two large captures and a small number of recaptured individuals. Fewer recaptures indicate a larger population. d. reflects the smallest population because few individuals were captured, and many were recaptured. Few captures and a high number of recaptures indicate a small population.

4. **Analyzing Data** Analyze the following situation. You are surveying two ponds, one large and one small, for goldfish. You catch, tag, and release 20 goldfish from each pond. The next day, you catch 20 goldfish from each pond and count 8 recaptures from the small pond and 2 from the large pond.

   a. Estimate the population of goldfish in the small pond.

   b. Estimate the population of goldfish in the large pond.

   c. Why would a large pond tend to have fewer recaptures than a small pond?

Answers may vary. Sample answers: a. $20 \times \frac{20}{8} = 50$**; b.** $20 \times \frac{20}{2} = 200$**;**

c. A large pond supports more fish, so there is a wider range of possible fish to recapture, reducing the chance of recapturing any one fish.

Name ______________________ Class ______________ Date ____________

Estimating Wild Animal Populations *continued*

5. **Analyzing Data** Analyze the following situation. You captured, marked, and released 5 turtles from a pond, and caught 10 unmarked turtles the next day. Would you have enough information to estimate the population using the mark-recapture method?

Answers may vary. Sample answer: No; the mark-recapture method requires at least one recaptured animal. Also, the sample is too small to get an accurate estimate of the population.

6. **Analyzing Results** Distinguish the following situation from the previous situation. If you captured and marked one turtle from a pond and captured the same turtle the next day, can you conclude that only one turtle lives in the pond? Explain your answer.

Answers may vary. Sample answer: According to the mark-recapture method, the estimated population in the pond is one turtle. However, there may be other turtles in the pond that were not captured. For the estimate to be accurate, there must be a larger number of turtles in the samples.

Conclusions

7. **Drawing Conclusions** Draw your conclusion on the use of the mark-recapture method for estimating a population. Is it a good way of estimating populations? Explain your answer.

Answers may vary. Sample answer: It is a good method for estimating, because if a large sample is taken both days, estimations can be reliable.

8. **Applying Conclusions** Imagine that you are studying birds that are flying south for the winter. How might their migration affect the results of a mark-recapture study? Can you accurately estimate the migrating bird population using the mark-recapture method? Explain your answer.

Answers may vary. Sample answer: If animals are migrating, the number of recaptures should be very low, making population estimates unrealistically high. It is difficult to accurately estimate the size of moving populations. The mark-recapture method works best with relatively stationary populations.

Skills Practice Lab

CBL™ PROBEWARE

Determining Growth Rate

Teacher Notes

Teacher Tested & Approved
Craig Kramer
Bexley High School
Bexley, Ohio

TIME REQUIRED One 45-minute period on the first day, 5–10 minutes per period on four to six observation days, 30 minutes on the final day

SKILLS ACQUIRED

Collecting data
Experimenting
Identifying and recognizing patterns
Interpreting
Measuring
Organizing and analyzing data

RATINGS Easy ← 1 2 3 4 → Hard

Teacher Prep–4
Student Setup–3
Concept Level–2
Cleanup–2

THE SCIENTIFIC METHOD

Make Observations In Procedure steps 13, 14, 16, 17, and 19, students make observations of growth medium samples.

Analyze the Results Analysis questions 4–6 and Conclusions question 1 require students to analyze their results.

Draw Conclusions Conclusions questions 4 and 5 ask students to draw conclusions from their data.

MATERIALS

The yeast solution is made by mixing 1 g active dry yeast per 100 mL of distilled water. Freeze-dried yeast or baker's yeast may be used.

Other juices could be substituted for apple juice, but select one that is clear. To be sterile, apple juice must remain unopened until the time of use. An alternative to apple juice as a growth medium is Sabouraud dextrose broth available from WARD'S. See the *Master Materials List* for ordering instructions.

SAFETY CAUTIONS

- Discuss all safety symbols and caution statements with students.
- Caution students not to taste any of the yeast solution or juice growth medium.

DISPOSAL

Juice and yeast solution may be poured down the drain. Destroy dry yeast by autoclaving or disinfecting with household bleach (5% sodium hypochlorite).

TECHNIQUES TO DEMONSTRATE

Demonstrate how to hold the colorimeter cuvette—by the top of the ribbed sides. This is the most secure way to hold the cuvette and minimizes chances for fingerprints on the smooth light-transmission surfaces.

TIPS AND TRICKS

Preparation

This lab works best in groups of two to four students.

The procedure in this lab is written for use with the original CBL system. If you are using CBL 2 or LabPro, the CHEMBIO program can still be used. Updated versions of this program can be downloaded from **www.vernier.com**. For additional information on how to integrate the CBL system into your laboratory, see the Program Introduction.

Some sensors may require the use of an adapter. Students will need to connect the adapter to the sensor before connecting it to the CBL.

To reduce observation time, calibrate the colorimeter and leave the CBL unit and calculator turned on during and between classes. Both the CBL unit and calculator will go into sleep mode. Press ON to turn them back on.

If you are using auto-ID colorimeters with CBL 2 or LabPro, the calibration instructions are slightly different from those in the lab procedure. Have students follow the on-screen instructions.

Yeast cell populations may be counted every other day and estimated from the final graph for days that were not counted. A correlation between the absorbance value and yeast cell population can be made.

Toward the end of the experiment, dead (lysed) yeast cell debris will be part of the mix, making the absorbance readings slightly higher than what would be expected with the cell count. For this reason, students are instructed in Analysis question 3 to disregard data once the growth rate is in decline, when making the graph in Figure 3.

Procedure

When students notice that the absorbance values have peaked and are in decline, typically in about 1 week, students can stop collecting data.

In step 7, students may use their growth medium (fruit juice) rather than distilled water for the colorimeter calibration. Try to use juice from the same container for consistency of color.

Students typically apply too much light when viewing yeast cells under a microscope, which makes the cells difficult to see. Yeast cells are most easily observed with low light levels. Stain the cells, if necessary, with Congo red.

Before removing a growth medium sample from the test tube, remind students to mix it well since the yeast cells tend to settle out.

As much as possible, readings should be taken at the same time each day.

The volume (in cm^3) of one field of view is calculated as follows: $\pi r^2 h$, where $\pi = 3.14$ and h is the height of the liquid between the slide and the coverslip, about 0.01 cm. Assume the volume to be 1.1×10^{-5} or 0.000011 mL.

Name ______________________ Class ______________ Date ____________

Skills Practice Lab

CBL™ PROBEWARE

Determining Growth Rate

Populations do not grow indefinitely. Their sizes fluctuate, and are determined by *emigration* (leaving a population), *immigration* (joining a new population), *natality* (births), and *mortality* (deaths). Changes in the environment and the availability of resources impact these factors.

In this lab, you will investigate a population of yeast. The population will be closed, and can change only by natality and mortality. You will use a colorimeter to determine how rapidly the yeast population grows. A colorimeter determines the concentration of a solution by measuring the amount of absorbed light from a beam shining through a sample. You will learn how to construct a standard curve that will allow you to determine the concentration of yeast cells/mL from the measured absorbance.

OBJECTIVES

Use a colorimeter to measure the light absorbance of a growth medium containing a population of yeast cells.

Graph a growth curve for a population of yeast cells.

Determine the concentration of yeast cells in your growth medium from a standard curve of absorbance versus yeast concentration.

Compare the yeast population growth curve to a logistic growth curve.

MATERIALS

- apple juice (growth medium)
- CBL System
- colorimeter cuvette with lid
- colorimeter
- coverslip
- culture tube, 20 × 150 mm and cap
- graduated cylinder, 10 mL
- graph paper
- lab apron
- link cable
- microscope
- microscope slide
- paper towel
- pipets, 15 cm disposable (2)
- safety goggles
- stopper, test-tube
- test tube, 15 mm × 125 mm (2)
- test-tube rack
- TI graphing calculator
- water, distilled
- yeast solution

Procedure

PREPARING THE GROWTH MEDIUM

1. Put on safety goggles and a lab apron.

2. Label a clean, dry 20 × 150 mm culture tube with your name and lab period. Add 30 mL of apple juice to the tube.

3. Obtain the yeast solution from your teacher. Swish the solution gently to mix it evenly. Then extract some into a pipet. Add 3 mL of the yeast solution to the apple juice in the tube. Pipet the medium gently up and down a few times to mix the juice and yeast solution. Leave your tube in a warm location, such as on a shelf.

SETTING UP THE CBL SYSTEM

4. Connect the CBL unit and the calculator with the link cable. Press the link cable firmly into each device to assure a good connection. Connect the colorimeter to Channel 1 of the CBL unit.

5. Turn on the CBL unit and the calculator. Start the CHEMBIO program. Go to the MAIN MENU.

6. Select SET UP PROBES. Enter "1" as the number of probes. Select COLORIMETER from the SELECT PROBE menu. Enter "1" as the channel number. You are now ready to set up the growth medium and calibrate the CBL unit and colorimeter.

CALIBRATING THE COLORIMETER

7. To avoid bubble formation on the cuvette sides, slowly fill a colorimeter cuvette $\frac{3}{4}$ full of distilled water. Wipe the outside dry with a paper towel. If bubbles are present, gently tap the cuvette with your finger to loosen them.

8. Place the cuvette into the colorimeter with the ribs facing the front and back of the colorimeter chamber, as shown in **Figure 1.** A smooth side of the cuvette should be facing the white reference mark on the colorimeter.

FIGURE 1 PLACE THE CUVETTE IN THE COLORIMETER

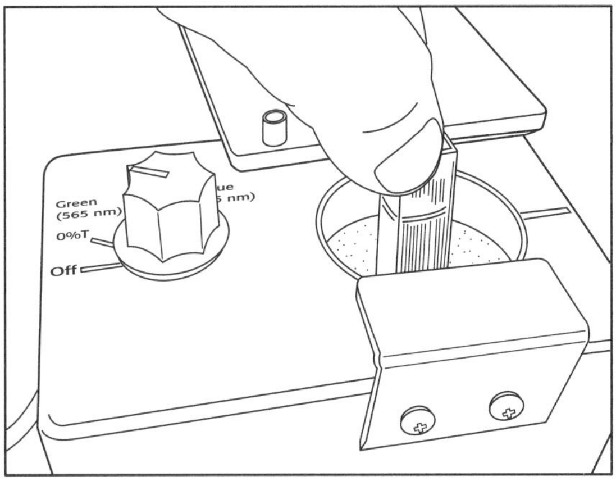

9. Close the colorimeter lid. Rotate the wavelength knob of the colorimeter to the "0 % T" position. When the display reading on the CBL unit is stable, press TRIGGER on the CBL unit.

10. On the graphing calculator, enter "0" as the reference. Turn the wavelength knob of the colorimeter to the Green LED position. When the CBL screen display is stable, press TRIGGER on the CBL unit. On the calculator, enter "100" as the reference. Leave the wavelength knob set to the Green LED setting to test samples.

11. On the calculator, press ENTER to return to the MAIN MENU. Select COLLECT DATA from the MAIN MENU. Select MONITOR INPUT from the DATA COLLECTION menu. On this setting, the CBL unit will monitor the light absorbance readings and display them on the graphing calculator.

MEASURING LIGHT ABSORBANCE

12. Agitate the culture tube gently to distribute the yeast organisms evenly. Remove the cap and withdraw about 2 mL of the growth medium with a sterile pipet. Fill a clean, dry cuvette with the growth medium from the pipet. Recap the culture tube immediately. Be sure the cuvette does not have any air bubbles in it. Wipe the outside dry with a paper towel. Rinse and dry the pipet.

13. Place the cuvette into the colorimeter. Close the lid and wait for the CBL display reading to stabilize. Read the absorbance value on the graphing calculator. If the absorbance reading is 1.0 or greater, follow the instructions in step 14 for diluting the growth medium sample. Record the absorbance reading and the dilution amount in **Table 1** for Day 1. If the growth medium was undiluted, record 1 for the dilution amount. If it was diluted by a factor of 10, record 0.1 for the dilution amount. If the first dilution was diluted, record 0.01 (1/100 dilution). Then, go to step 15.

14. Dilute your growth medium sample if the absorbance value on the colorimeter is greater than 1.0. Into a clean, dry test tube, place 1 mL of the growth medium from the cuvette. To the test tube containing the 1 mL of growth medium, add 9 mL of fresh juice. This produces a dilution of $\frac{1}{10}$ of the sample. Stopper the test tube, and shake gently to mix the contents. Remove 2 mL and place it into a clean cuvette. Wipe the cuvette and place it into the colorimeter. Follow the instructions in step 13 to measure the absorbance.

15. Turn off the CBL System by pressing "+" on the calculator, then QUIT. Remove the cuvette from the colorimeter, and save the solution to measure yeast concentration.

Name ______________________ Class ______________ Date ____________

Determining Growth Rate *continued*

MEASURING THE YEAST CONCENTRATION

16. Place a lid on the cuvette, and shake it gently to mix the medium evenly. Using a clean pipet, remove some medium from the cuvette. Place two drops on a microscope slide. Cover the drops with a clean coverslip. Put the slide on a microscope. Focus on low power, change to high power, and refocus as necessary. Adjust the light to view the yeast cells most easily.

17. Count the number of yeast cells visible in the high power (HP) field of view. A yeast cell with a bud counts as two. Record the number of yeast cells in **Table 1** under "Yeast counted." If time permits or if your teacher directs you to do so, move the slide to observe another section of it, or make a second slide. Then make another HP field count of yeast cells. Average the two counts, and record the average in **Table 1** under "Yeast counted."

18. Clean and dry the microscope slide and coverslip. Empty the cuvette into the sink, then rinse, and dry the cuvette. Clean up your work area and wash your hands. Store your labeled culture tube in a test-tube rack in a warm location.

19. Repeat steps 4–18 each day for up to 8 days or until you notice the absorbance values have peaked and are in decline.

TABLE 1 DAILY ABSORBANCE AND YEAST CONCENTRATION MEASUREMENTS

Day	Measured absorbance (%)	Dilution	Absorbance (%)	Yeast counted (no. of cells)	Actual yeast (no. of cells)	Cell concentration (no./mL)
1	**0.264**	**1**	**0.264**	**410**	**410**	**3.72×10^7**
2	**0.584**	**1**	**0.584**	**712**	**712**	**6.47×10^7**
3	**0.821**	**1**	**0.821**	**760**	**760**	**6.9×10^7**
4	**0.121**	**0.1**	**1.21**	**90**	**900**	**8.18×10^7**
5	**0.239**	**0.1**	**2.39**	**228**	**2280**	**2.07×10^8**
6	**0.516**	**0.1**	**5.16**	**400**	**4000**	**3.64×10^8**
7	**0.323**	**0.1**	**3.23**	**224**	**2240**	**2.04×10^8**
8	**0.310**	**0.1**	**3.10**	**144**	**1440**	**1.31×10^8**
9	**0.287**	**0.1**	**2.87**	**80**	**800**	**7.27×10^7**

Data will vary. Sample data are entered above.

Analysis

1. **Organizing Data** Calculate the absorbance and actual yeast population in a high power field for each day. Record your values in **Table 1.** Compute absorbance by dividing the measured absorbance in **Table 1** by the dilution. If the dilution is 1.0, then the absorbance value will equal the measured absorbance value.

   Example: If the measured absorbance is 0.121 and the dilution is 0.1,

$$\frac{\text{measured absorbance}}{\text{dilution}} = \frac{0.121}{0.1} = 1.21$$

   Compute the actual yeast population by dividing the diluted yeast population by the dilution.

   Example: If the diluted yeast population is 90 and the dilution is 0.1,

$$\text{actual yeast population} = \frac{\text{diluted yeast population}}{\text{dilution}} = \frac{90}{0.1} = 900$$

2. **Organizing Data** Calculate cell concentration, using the data from **Table 1.** The cell concentration can be determined as follows:

$$\text{no. yeast cells/mL} = \frac{\text{(actual number of cells in one HP field of view)}}{0.000011\ \text{mL}}$$

   Record the values of cell concentration in **Table 1.** (0.000011 mL is the volume in one field of view)

3. **Constructing Graphs** Use the data in **Table 1** to construct two graphs. In **Figure 2,** graph absorbance (y-axis) against the number of days (x-axis). In **Figure 3,** graph yeast cell concentration (y-axis) against absorbance (x-axis). Only use data prior to the decline of the growth rate. Draw a line of best fit in **Figure 3.**

FIGURE 2 ABSORBANCE VERSUS TIME

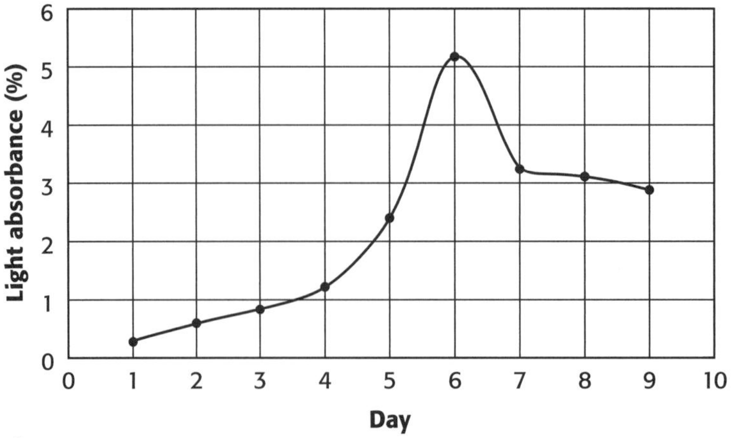

Sample graph

Name ______________________________ Class ______________ Date ____________

FIGURE 3 YEAST CONCENTRATION VERSUS ABSORBANCE

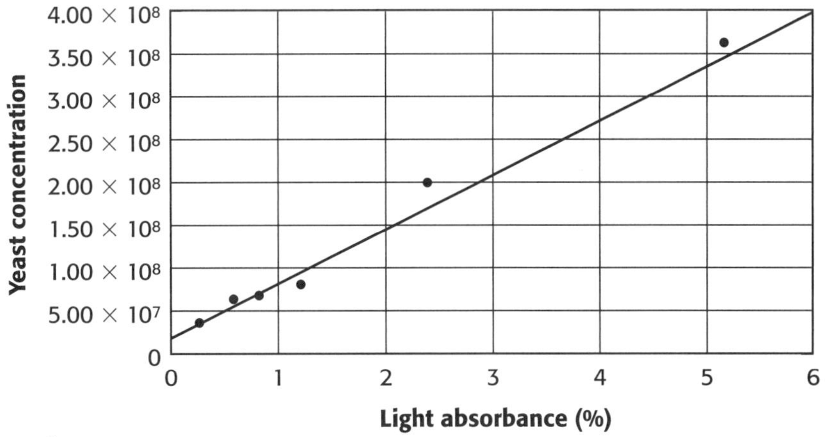

Sample graph

4. **Analyzing Graphs** Look at your graph of absorbance versus time. Describe the growth pattern of yeast during the experiment.

   Answers will vary. In general, students should observe a rapid increase in light absorbance and yeast concentration followed by a sharp decline.

5. **Analyzing Graphs** The concentration of growing yeast is important to biologists because they often need to use the cells while they are still growing exponentially. Look at your graph of absorbance versus time. During what range of absorbance are these cells growing exponentially?

   Answers will vary. Results should be consistent with students' graphs. According to the sample data, cells are growing exponentially in the light absorbance range from about 2–5 %.

6. **Identifying Relationships** Biologists often use absorbance to determine the concentration of yeast and bacteria rather that doing actual cell counts each time. By using a standard curve such as the one you constructed in **Figure 3**, biologists can easily determine the concentration of microorganisms in growth media. What was the concentration of yeast in your sample when the absorbance was 1%? 2%?

   According to the sample data, when the absorbance is 1%, the concentration is about 800 cells/mL. At an absorbance of 2%, the concentration is about 1400 cells/mL.

Name ______________________ Class ______________ Date ____________

Conclusions

1. **Analyzing Graphs** Suppose you want to collect the most yeast you can for an experiment in which the yeast need to be growing exponentially. At which absorbance would you collect your yeast?

   Answers will vary according to students' graphs. Sample data indicate that to collect the maximum amount of yeast, collect when the light absorbance reaches 5%.

2. **Evaluating Methods** During the dilution process, why was juice used rather than water?

   The juice matches the growth medium. The optical property (absorbance) of the juice will be the same as the growth medium.

3. **Evaluating Results** If there are no limits to the growth of a population, the population will grow exponentially. But because there are limited resources in nature, a population's growth will slow, and fluctuate around a number called its *carrying capacity*. The carrying capacity is the maximum population size an environment can support for a long period of time. This kind of growth is called logistic growth, and is modeled by the *logistic growth curve*, shown in **Figure 4.** Does your graph in **Figure 2** look like the logistic model? In what ways is it similar or different?

   The graph will probably not look like the logistic model. Students' graphs will most likely show an exponential rise and then will drop off steeply. The graphs should be similar to the logistic model in the exponential phase, but different from the logistic model in that the graphs continue to drop somewhat instead of truly leveling off at a carrying capacity.

FIGURE 4 THE LOGISTIC POPULATION GROWTH MODEL

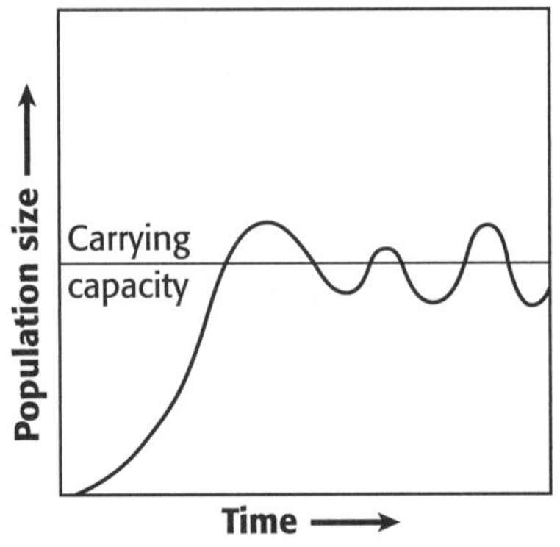

Name ______________________ Class ______________ Date ____________

Determining Growth Rate *continued*

4. **Drawing Conclusions** Why do you think your graph in Figure 2 does not level off at, or fluctuate around, a carrying capacity? (Hint: Suppose that the graph in **Figure 4** represents the population growth of moose on an island. Consider the differences between the growth of the moose population on the island and the yeast population in the media when accounting for any differences in the graphs.)

   Answers will vary, but students should understand that, unlike the food supply for the moose population, the food supply for the yeast population was not replenished. This helps lead to a continued decline of the yeast population rather than a steadying of the population around a carrying capacity.

5. **Drawing Conclusions** Describe the factors that contributed to the yeast population growth and decline.

   The growth of the yeast was due to a plentiful energy source and space to grow. As the population became more crowded, food became scarce, waste and cellular debris accumulated, and space was taken up. Because no further food was provided, the population crashed.

Extensions

1. Design an experiment to test and investigate the effects of other growth mediums on yeast growth. Other clear juices, sugar solutions, or Sabouraud dextrose broth may be good choices for other growth mediums.
2. Use the library, media center, or Internet to research the growth of the human population. Find out when and why the human population began to grow rapidly. Learn about the past and current human population sizes and draw a graph of human population growth. Compare it to the logistic model. Develop a visual presentation for your class. Include information about population growth in developed and developing countries, and human carrying capacity.

Inquiry Lab

CONSUMER

Bug Off!

Teacher Notes

TIME REQUIRED Two 45-minute class periods

SKILLS ACQUIRED

Collecting data
Communicating
Constructing models
Designing experiments
Experimenting
Interpreting
Organizing and analyzing data
Predicting

RATING

Easy ← 1 2 3 4 → Hard

Teacher Prep–1
Student Set-Up–2
Concept Level–3
Clean Up–2

THE SCIENTIFIC METHOD

Ask Questions Procedure, step 1

Test the Hypothesis Analysis, question 1

Analyze the Results Analysis, questions 1 and 2

Communicate the Results Conclusions, question 4

MATERIALS

Sources for obtaining ants as well as recipes for ant repellents can be found primarily on the Internet. Ants should be ordered by the teacher. Some states require a USDA permit prior to shipping ants.

SAFETY CAUTIONS

Some students are nervous around insects. Students should refrain from making sudden, startled movements, to prevent injuries. Animals used in research labs must be treated humanely and respectfully. Playing with them should not be allowed. Use caution and make sure that all students are wearing the appropriate laboratory protections.

Beware of possible allergic reaction to ants, plants, or other materials used in this lab. Excuse students from handling ants or materials if they have any known allergies. Be sure that spilled or dropped materials are cleaned up immediately in order to avoid accidental slips and falls.

You may have to order a Federal permit to use ants in the classroom, depending on the species.

DISPOSAL

Nonreproductive harvester ants are preferable for this experiment because they are ideal for placing in ant farms after the laboratory.

CHECKPOINTS

Checkpoint 1 By the end of the first class period, students should turn in a detailed, one-page procedure for approval. The procedure should cover each step of the scientific method and should ensure that each of the objectives is met. Students should also have constructed their cardboard model to test the effectiveness of keeping ants out of a home.

Checkpoint 2 By the end of the second class period, students should have mixed their repellent and performed the entire procedure.

TIPS AND TRICKS

The experiment and model that the students design should be small in scale. This will make it easier for them to carry out the experiment and model in the time provided. Find out the favorite food of the ants used in the experiment. This can be a good "bait" to ensure a successful lab.

Name ______________________________ Class ______________ Date ____________

Inquiry Lab

CONSUMER

Bug Off!

You are the lead research scientist for Bug Off! Corporation, the world's leading producer of insect repellents, and you have just received the following memo:

To: Director of Research and Development

From: Director of Marketing

cc: Director of Production

Subject: NEED FOR A NEW CONSUMER REPELLENT

We are getting reports from many states that the ant population is soaring and is causing many problems to homeowners. Existing ant repellents are not effective enough. We have an opportunity to increase our market share if we are first to the store shelves with a new product. Information I have says that Insects Be Gone!, our major competitor, is two weeks from production of their new product. Bug Off! needs to produce an effective, inexpensive and environmentally friendly ant repellent. During your research, please remember that every step of the scientific method must be considered and adapted to your development efforts.

I also urge you to use only common, readily accessible ingredients. This keeps the cost of materials down. The success of our efforts will also be determined by our decision to use nonhazardous, environmentally safe materials that help keep the cost of production low and protect the environment. Good luck!

OBJECTIVES

Collect data on the effects of your newly developed repellent on individual ants.

Construct and **test** a model of a home being invaded by ants to determine if the repellent is effective.

MATERIALS

- ants
- graph paper
- modeling supplies, common (cardboard, glue, adhesive tape, etc.)
- natural, nonhazardous ant repellent materials, such as peppermint, garlic, ground cinnamon, or spearmint plants
- standard lab equipment for measuring and mixing

Procedure

1. Brainstorm with your lab team and develop a procedure that uses each of the steps of the scientific method and helps you meet your objectives. After you have designed the procedure, write the steps in the space provided below. The questions that start on the next page allow you to present key pieces of information that you should learn while following your procedure.

   Answers may vary.

Analysis

1. **Constructing Graphs** Create a table, chart, or graph to document the data you collected on the effectiveness of the repellent on the ants. Present your table, chart, or graph on a separate sheet of graph paper.

   Answers may vary based upon the procedure developed by the student.

2. **Analyzing Results** What percentage of individual ants were repelled by your product? Explain your analysis.

   Answers may vary based upon the procedure developed by the student.

Conclusions

3. **Analyzing Graphs** Explain the data that shows the ability of your product to keep ants away. If the repellent worked on ants in the laboratory, will it work in a real life situation? Explain your analysis.

   Answers may vary based upon the procedure developed by the student.

4. Evaluating Models Evaluate the effectiveness of your cardboard model. Use the space provided below to explain the model you used in pictures and words. Write the explanation on the lines provided below the space.

Answers may vary based upon the procedure developed by the student.

Extension

1. Building Models How effective would your repellent be if the ants were twice as big as those you used in your experiment and model?

Answers may vary based upon the procedure developed by the student.

Exploration Lab

LONG-TERM PROJECT

Calculating Generation Rate

Teacher Notes

TIME REQUIRED Five 45-minute class periods

Alonda Droege
Evergreen
High School
Seattle, Washington

SKILLS ACQUIRED

Collecting data
Communicating
Designing experiments
Experimenting
Interpreting
Organizing and analyzing data
Predicting

RATING Easy ← 1 2 3 4 → Hard

Teacher Prep–3
Student Set-Up–2
Concept Level–2
Clean Up–2

THE SCIENTIFIC METHOD

Make Observations Procedure, steps 1–10

Form and Test the Hypothesis Extension, question 1

Analyze the Results Analysis, questions 1 and 2

Draw Conclusions Conclusions, questions 3 and 4

MATERIALS

Prepare in advance separate stocks of *Blepharisma* and *Euplotes* microorganisms using bottled spring water and boiled wheat germ kernel in clean glass bowls. Sources for obtaining microorganisms and instructions for reproduction can be found on the Internet. Using sterile Petri dishes with dividers for culture vessels will help to avoid contaminants.

SAFETY CAUTIONS

Ensure that students do not drink the microorganism cultures and that they wear the appropriate lab clothing and eye protection.

DISPOSAL

Dump the contents and rinse all of the culture dishes and stocks that are no longer in use into a common collection container. When the lab has been completed, this liquid waste should be boiled to kill the microorganisms.

Name ______________________ Class ______________ Date ____________

Exploration Lab

LONG-TERM PROJECT

Calculating Generation Rate

Organisms compete for resources as they interact with their environment. Competition is a relationship in which different individuals or populations attempt to use the same limited resource. A resource is that which an organism needs to support itself to maintain its population. Resources include foods, open spaces, and hiding places. When organisms compete for resources, populations may be limited. Conditions are not resources, but are factors that may affect growth and reproduction.

Russian ecologist G. F. Gause conducted early experiments on competition. He grew *Paramecium aurelia* and *Paramecium caudatum* alone and together and provided them with food resources. Each species grew well alone, but when placed together, only *P. aurelia* survived. Gause then changed the conditions by adding sediment to the environment. The new conditions allowed *P. caudatum* to take refuge in the sediment. *P. caudatum* continued to multiply in this hiding place and coexisted with *P. aurelia*. Hiding in the sediment is an example of an adaptation to changing environmental conditions.

Results from many studies were summarized into what is known as the competitive exclusion principle. This principle states that two species cannot coexist indefinitely on the same limited resource. Organisms that are more adaptive tend to grow and displace those that do not adapt as well to changing conditions.

An interesting measure of growth is the generation rate. This is the number of generations of an organism that have been produced during a specified amount of time. The number of generations that have occurred, also called the generation number, can be calculated using the equation:

$$\text{Generation number (g)} = \frac{(\log B - \log A)}{\log 2}$$

"A" is a count of the cells at one time, and "B" is the count of the cells at a later time. The equation calculates the number of generations "g" that have occurred between the two times. Divide "g" by the elapsed time between the two counts, and you have determined the generation rate. Organisms with larger generation rates can overwhelm and outcompete other organisms.

In this lab, you will gather data and plot growth curves of noncompeting and competing populations of two types of organisms. You will analyze your data by comparing generation rates of the various populations.

OBJECTIVES

Experiment on organism competition.

Predict the effects of competition on number differences between competing populations.

Design an experiment to test the effects of competition on changing an environmental condition.

Name ______________________ Class ______________ Date ____________

MATERIALS

- culture dish, six-well plastic with cover (3)
- distilled water, boiled and cooled (stock)
- microorganisms, *Blepharisma* (stock in glass bowl)
- microorganisms, *Euplotes* (stock in glass bowl)
- microscope, stereoscopic dissecting
- micropipettes, Pasteur, glass, fine-tipped, fitted with rubber bulbs
- paper, graph (2)
- water, bottled spring (stock)
- wheat germ kernel

Procedure

1. Label each well of a six-well plastic culture dish with the numbers 1 through 6. Fill each of the wells of the culture dish with 5 mL of bottled spring water. Add 0.5 mL of wheat germ kernel to each well.
2. Place one *Blepharisma* in each well. Place covers on each of the six wells and label the six-well plastic culture dish "*Blepharisma* Only."
3. Repeat step 1 using a separate six-well plastic culture dish. Place one *Euplotes* in each well. Place covers on each of the six wells and label this culture dish "*Euplotes* Only." Before beginning the experiment, make a hypothesis and record it below.

   ____

   ____

4. Count the number of cells per well daily for five days. Perform the count using a stereoscopic dissecting microscope. With some practice, you should be able to estimate the number of cells for each genus by scanning the contents of each well. Slow-moving ciliates such as *Blepharisma* are easy to count by eye. You can also use the micropipettes to do manual counts. Remove each organism as it is counted to another container while viewing through the stereoscopic dissecting microscope. When finished counting a well, be sure to return all organisms to the well from which you removed them.
5. Record the results of the counts of step 4 in Table 1.
6. Repeat step 1 using a separate six-well plastic culture dish. Place one individual cell of each of the two genera into each of wells 1 through 4.
7. Place two cells of the *Blepharisma* in well 5. This is a control population for the *Blepharisma*.

Name ______________________ Class ______________ Date ____________

Calculating Generation Rate *continued*

TABLE 1: *BLEPHARISMA* ENVIRONMENT POPULATION COUNTS

	1 day	2 days	3 days	4 days	5 days
Well 1					
Well 2					
Well 3					
Well 4					
Well 5					
Well 6					

TABLE 2: *EUPLOTES* ENVIRONMENT POPULATION COUNTS

	1 day	2 days	3 days	4 days	5 days
Well 1					
Well 2					
Well 3					
Well 4					
Well 5					
Well 6					

TABLE 3: MIXED/CONTROL ENVIRONMENT POPULATION COUNTS

	1 day	2 days	3 days	4 days	5 days
Well 1 ***Blepharisma***					
Well 1 ***Euplotes***					
Well 2 ***Blepharisma***					
Well 2 ***Euplotes***					
Well 3 ***Blepharisma***					
Well 3 ***Euplotes***					
Well 4 ***Blepharisma***					
Well 4 ***Euplotes***					
Well 5 ***Blepharisma***					
Well 5 ***Euplotes***					
Well 6 ***Blepharisma***					
Well 6 ***Euplotes***					

8. Place two cells of the *Euplotes* in well 6. This is the control population for the *Euplotes*.
9. Place covers on each of the six wells and label this culture dish "Mixed/Control."
10. Count the number of cells per well daily for five days. Follow the instructions of step 4 on how to perform a count. Record the results in Table 2.

Analysis

1. **Constructing Graphs** Use a separate sheet of graph paper to plot growth curves for the *Blepharisma* populations. Place the title "*Blepharisma* Populations" at the top of the graph paper. Using data in Table 1, plot the average number of *Blepharisma* you observe each day on the y-axis and the number of days on the x-axis. Label the curve "No competition." On this same graph, use the data in Table 3 to plot a curve of the average number of *Blepharisma* you observe each day in the mixed environments. Label this curve "Competition." On this same graph, use the data in Table 3 to plot a curve of the number of *Blepharisma* you observe each day in the control environment. Label this curve "Control." Use the data in Table 2 and Table 3 and another piece of graph paper to plot growth curves for *Euplotes* populations. All averages are to be calculated by adding the counts in each well for a given day and dividing by the number of wells.
Answers may vary. Students should have two separate graph papers.

2. **Organizing Data** Calculate a generation rate for each of the six curves on your graphs. Show all calculations.

Answers may vary. There should be six calculations. Sample calculation:

g of mixed *Blepharisma* $= \dfrac{(\log 20 - \log 2)}{\log 2}$ **= 3 generations; t = 1 day;**

$\dfrac{g}{t}$ **of mixed *Blepharisma*** $= \dfrac{3 \text{ generations}}{1 \text{ day}}$ **= 3 generations per day**

Conclusions

3. **Drawing Conclusions** Draw conclusions as to which organism will outcompete the other in the mixed environment. What do the generation rates for each of the six population curves suggest about this competition? How does this compare to your hypothesis?

Answers may vary. Sample answer: The generation rates for the control populations are the same as for single organism populations. This shows that I can compare the single organism populations directly to the mixed populations. The single organism population with the largest generation rate should win in a competition. This is verified in the mixed population generation rates.

4. **Making Predictions** Imagine that you created a mixed environment starting with five times more of the organism with the lower generation rate. Do you think this would change the results of a competition? Explain.

Answers may vary. Sample answer: I think the organism with the lower generation rate would win. The generation rates are close together, so giving one microorganism a head start would allow it to occupy space and take control of resources faster than the other microorganism.

Extension

1. **Designing Experiments** Design an experiment (objective, materials, and procedure) that tests the effect of a change in environmental conditions on the two microorganisms. Decide which condition you will change compared to the experiment you already performed. You could use G. F. Gause's experiment as an example. Make sure that you include unmixed, mixed, and control populations in your experiment.
Answers may vary. The experiment design should fill one page.

Answer Key

Concept Review

MATCHING

1. h
2. f
3. a
4. j
5. d
6. i
7. b
8. c
9. e
10. g

MULTIPLE CHOICE

11. c
12. a
13. d
14. b
15. a
16. b
17. a
18. b
19. d
20. b
21. b
22. a

Critical Thinking

ANALOGIES

1. b
2. d
3. c
4. b
5. c
6. b
7. b
8. c

INTERPRETING OBSERVATIONS

9. The arboreal monkey population had better access to two readily available resources (fruits and leaves), but the terrestrial monkeys depended on a few small insects and fruits falling from the trees. Their resources could not support a bigger population.
10. Competition. The arboreal population had advantageous access to the same fruits consumed by the terrestrial monkeys.
11. No. It is surviving and has access to a second food resource (insects) that can increase its chance for continuous reproduction. This may allow the species to adapt to new resources found on the ground and eliminate its competition for fruit with the arboreal monkeys.

AGREE OR DISAGREE

12. Disagree; the organisms that live in an ecosystem are integral parts of that ecosystem, while parasites are not integral parts of their hosts. Also, the organisms in an ecosystem do not usually harm the ecosystem. However, parasites live on (or in) living hosts that they do harm.
13. Disagree; rodents, deer, raccoons, and other organisms also compete with humans for food.
14. Disagree; reintroducing the gray wolf can generate several types of interactions. Hunters of wild moose, deer and elk feel that wolves reduce the available population of game animals. This interaction can be defined as consumption. National Parks and Natural Reserves favor reintroducing wolves because revenues from hunting fees help support maintenance and research programs. This interaction can be defined as mutualism.

REFINING CONCEPTS

15. No; within any ecosystem, each species plays a distinct role by which it gathers the resources necessary for life. Since resources such as sunlight, food, and water are limited, organisms that attempt to use those resources in exactly the same way will compete. Competition will lead species to evolve different strategies for acquiring resources and, therefore, different niches within their environments. Although two species can have niches that overlap, their niches are not the same.
16. Yes; parasites rob their hosts of energy and nutrients. For example, ticks suck blood, and tapeworms eat food that the host is digesting. Hosts with adaptations that reduce the impact of their parasites will have a better chance of surviving and reproducing (natural selection). These adaptations will be passed on to the host's offspring and eventually change the genetic characteristics of the host's population.

Evolution is a change in the genetic characteristics of a population, so the parasite would have influenced the evolution of its host species through the process of natural selection.

17. Yes, roses mate with the help of bees and other flying insects that transfer pollen from plant to plant as they search for nectar. Since flying insects can easily cross a wide road, a rose on one side of the road has a reasonable chance of mating with one on the other side and is therefore part of the same population.

Active Reading

SECTION: HOW POPULATIONS CHANGE IN SIZE

1. b
2. c
3. a
4. death rate
5. The average number of births would equal the average number of deaths, so the growth rate would be zero. If half the population has two offspring *each* and the other half has none, the birth rate (0.5 x 2 = 1.0) is the same as if every *pair* of adults had two offspring each. If all members of the adult population then die, the growth rate remains zero.
6. 4
7. 1
8. 2
9. 3
10. Negative growth rate is when a population's average number of deaths is greater than its average number of births. Zero growth rate is when the average number of deaths equals the average number of births.
11. In both negative and zero growth rate, the average number of births does not exceed the average number of deaths.
12. The population would decrease.

SECTION: HOW SPECIES INTERACT WITH EACH OTHER

1. Students may choose four of the following: ticks, fleas, tapeworms, heartworms, bloodsucking leeches, and mistletoe.
2. nourishment
3. parasitism
4. a
5. b
6. an organism alongside its food
7. the practice of being alongside the food
8. Parasites and predators both depend on other organisms for survival.
9. A parasite spends some of its life in or on its host; parasites do not usually kill their hosts.
10. The longer a host lives, the longer a parasite will have a source of nourishment.
11. The host can be weakened or exposed to disease when it carries a parasite.

Map Skills

1. 1953
2. 50 years
3. perch
4. It will decrease again by 50 percent.

Quiz

SECTION: HOW POPULATIONS CHANGE IN SIZE

Matching	Multiple Choice
1. c	6. a
2. d	7. d
3. b	8. c
4. e	9. b
5. a	10. d

SECTION: HOW SPECIES INTERACT WITH EACH OTHER

Matching	Multiple Choice
1. b	6. b
2. a	7. d
3. d	8. c
4. e	9. a
5. c	10. d

Chapter Test General

MATCHING

1. h	6. b
2. d	7. f
3. a	8. e
4. j	9. c
5. i	10. g

MULTIPLE CHOICE

11. a
12. c
13. a
14. b
15. d
16. d
17. c
18. b
19. c
20. a

Chapter Test Advanced

MATCHING

1. i
2. f
3. a
4. j
5. b
6. c
7. g
8. d
9. h
10. e

MULTIPLE CHOICE

11. c
12. b
13. d
14. b
15. b
16. d
17. d
18. a
19. c
20. c

SHORT ANSWER

21. The relationship between the bats and cacti is mutualism; the bats eat the nectar in the cacti's flowers and spread pollen and seeds. With a decreased bat population, flowers will go unpollinated and fruit will go uneaten, reducing the number of opportunites for the cactus plants to reproduce.

22. Answers may vary. Sample answer: The population of coyotes might also experience exponential growth. The coyotes are predators of rabbits and would have an abundant source of food as the rabbit population grows increasingly faster. With plenty of food available, more coyotes would survive to reproduce.

23. Answers may vary. Sample answer: As the rabbits exceed the carrying capacity, they will deplete resources. Starvation and possibly disease will severely reduce the population of rabbits. The coyotes will have less food available and their population will decline unless they find another food source.

24. Answers may vary. Sample answer: A species of plant with red, tubular flowers and a species of hummingbird with long beaks may have coevolved. The hummingbird benefits by being attracted to the flowers that provide food. The plant benefits when the hummingbird pollinates the flowers it visits.

ALTERNATIVE ASSESSMENT

25. a. mutualism
b. competition
c. commensalism
d. predation or parasitism

TEST ITEM LISTING

Understanding Populations

MULTIPLE CHOICE

1. Thick fur on deer is *not* an example of coevolution. Why?
   a. because thick fur is an adaptation
   b. because deer with thick fur live longer
   c. because thick fur evolved in response to a cold climate, not in response to other organisms
   d. because in the lowlands, where the climate was sunny and warm, deer that did not have thick fur became separated from other deer that did have thick fur

   Answer: c　Difficulty: 2　Section: 2　Objective: 5

2. An example of a population is
   a. all trees in a forest.
   b. all maple trees in a forest.
   c. all plants in a forest.
   d. all animals in a forest.

   Answer: b　Difficulty: 1　Section: 1　Objective: 1

3. The density of a population is
   a. the number of individuals born every year.
   b. the proportion of males and females.
   c. the number of individuals living in cities.
   d. the number of individuals per unit area.

   Answer: d　Difficulty: 1　Section: 1　Objective: 1

4. Each of the following is an example of a parasite *except*
   a. a roundworm in a human's intestine.
   b. a cow in a pasture.
   c. a tick on a cat.
   d. mistletoe on a tree.

   Answer: b　Difficulty: 1　Section: 2　Objective: 3

5. The relationship between a Canadian lynx and a snowshoe hare is an example of
   a. parasite and host.
   b. predator and prey.
   c. competition.
   d. mutualism.

   Answer: b　Difficulty: 1　Section: 2　Objective: 3

6. In which of the following relationships is neither species harmed?
   a. predation
   b. competition
   c. parasitism
   d. commensalism

   Answer: d　Difficulty: 1　Section: 2　Objective: 3

7. Which of the following populations has a random dispersion?
   a. flock of flamingoes
   b. pine trees in a pine forest
   c. herd of bison
   d. solitary snakes in a desert

   Answer: d　Difficulty: 1　Section: 1　Objective: 1

8. Which of the following would most likely cause a large number of density-independent deaths in a population?
   a. winter storms
   b. disease-carrying insects
   c. predators
   d. limited resources

   Answer: a　Difficulty: 1　Section: 1　Objective: 4

9. Which of the following organisms has the highest reproductive potential?
   a. dogs
   b. elephants
   c. bacteria
   d. humans

   Answer: c　Difficulty: 1　Section: 1　Objective: 3

10. A species of plant has exponential growth after it is introduced into an area where it has never lived. Which statement best describes exponential growth?
    a. Each individual plant grows much larger than usual.
    b. The population immediately decreases.
    c. Within a few years the population increases dramatically.
    d. The species' reproductive potential declines.

    Answer: c Difficulty: 1 Section: 1 Objective: 2

11. The relationship between acacia trees and the ants that live on them is an example of
    a. commensalism.
    b. mutualism.
    c. parasitism.
    d. predation.

    Answer: b Difficulty: 1 Section: 2 Objective: 3

12. The number of wild horses per square kilometer in a prairie is the horse populations
    a. density.
    b. dispersion.
    c. size.
    d. birth rate.

    Answer: a Difficulty: 1 Section: 1 Objective: 1

13. A female dog's niche includes all of the following *except*
    a. fleas that infest the dog.
    b. the number of puppies the dog has.
    c. how the dog protects its owners.
    d. the neighbor's enclosed yard.

    Answer: d Difficulty: 1 Section: 2 Objective: 2

14. If over a long period of time, each pair of adults in a population had only two offspring and the offspring lived to reproduce, the population would
    a. grow.
    b. shrink.
    c. remain the same.
    d. disperse randomly.

    Answer: c Difficulty: 1 Section: 1 Objective: 3

15. Which of the following has the greatest effect on reproductive potential?
    a. producing more offspring at a time
    b. reproducing more often
    c. having a longer life span
    d. reproducing earlier in life

    Answer: d Difficulty: 1 Section: 1 Objective: 3

16. A true statement about parasitism is that parasites
    a. may cause their hosts to become more vulnerable to predators.
    b. do not live on or in their hosts' bodies.
    c. are always animals and never plants.
    d. immediately kill their hosts.

    Answer: a Difficulty: 2 Section: 2 Objective: 4

17. Which of the following is *not* an example of exponential growth?
    a. rabbit populations after being introduced to Australia
    b. reindeer of the Probilof Islands after eating most of the lichens
    c. a bank account that earns interest
    d. mold appearing on bread overnight

    Answer: b Difficulty: 1 Section: 1 Objective: 2

18. The carrying capacity of an environment for a particular species at a particular time is determined by the
    a. number of individuals in the species.
    b. distribution of the population.
    c. reproductive potential of the species.
    d. supply of the most limited resources.

    Answer: d Difficulty: 1 Section: 1 Objective: 4

19. Which of the following statements explains why the growth of orchids on the high branches of tropical trees is an example of commensalism?
   a. The orchids draw nourishment from the trees.
   b. The trees are neither benefited nor harmed.
   c. The orchids keep parasites away.
   d. The trees receive nutrients from the orchids.

   Answer: b Difficulty: 1 Section: 2 Objective: 3

20. Which of the following statements is *not* correct?
   a. Mutualism is a type of symbiosis.
   b. Yucca moths and yucca plants have a symbiotic relationship.
   c. Symbiosis is a relationship in which two organisms live apart.
   d. Coyotes and foxes are competitors because they feed on the same kinds of animals.

   Answer: c Difficulty: 1 Section: 2 Objective: 5

21. Which of the following is one of the main properties used to describe a population?
   a. number of individuals
   b. color of individuals
   c. number of species
   d. kind of adaptations

   Answer: a Difficulty: 1 Section: 1 Objective: 1

22. Which of the following statements is correct?
   a. An organism's niche is only the part of its habitat that it eats.
   b. An organism's habitat is a location.
   c. Habitat and niche are the same thing.
   d. An organism's niche is outside its habitat.

   Answer: b Difficulty: 1 Section: 2 Objective: 1

23. Competition for food *cannot* occur
   a. between two populations.
   b. among members of the same population.
   c. among populations whose niches overlap.
   d. between animals from two different ecosystems.

   Answer: d Difficulty: 1 Section: 2 Objective: 3

24. Which of the following reproductive situations will limit a population's biotic potential?
   a. the minimum number of offspring each pair can produce
   b. the maximum number of offspring each individual can produce
   c. the number of interactions each individual has
   d. the size of offspring each individual can produce

   Answer: b Difficulty: 1 Section: 1 Objective: 3

25. The difference between a predator and a parasite is that a predator
   a. usually kills and eats its prey.
   b. benefits from another organism.
   c. lives in or on a host.
   d. harms another organism.

   Answer: a Difficulty: 1 Section: 2 Objective: 4

COMPLETION

26. A population's ____________________ is usually described as even, clumped, or random.

   Answer: dispersion Difficulty: 2 Section: 1 Objective: 1

27. A robin that does not affect the tree in which it nests is an example of ____________________.

   Answer: commensalism
   Difficulty: 2 Section: 2 Objective: 3

28. If two species use the same food source at different times, they are ____________________ competitors.

    Answer: indirect Difficulty: 2 Section: 2 Objective: 3

29. Unlike a predator in relation to its prey, a parasite does not usually ____________________ its host.

    Answer: kill and eat
    Difficulty: 2 Section: 2 Objective: 4

30. The average age at which members of a species reproduce is that species' ____________________.

    Answer: generation time
    Difficulty: 2 Section: 1 Objective: 3

31. The maximum number of offspring that each member of a population can produce is the population's ____________________.

    Answer: reproductive potential
    Difficulty: 2 Section: 1 Objective: 3

32. The three main properties used to describe a population are ____________________, ____________________, and ____________________.

    Answer: size, density, dispersion
    Difficulty: 2 Section: 1 Objective: 1

33. The ____________________ of an ecosystem for a particular species is the maximum population that the ecosystem can support indefinitely.

    Answer: carrying capacity
    Difficulty: 2 Section: 1 Objective: 4

34. The amount of food available for wolves in an area determines the ecosystem's carrying capacity for wolves and is a(n) ____________________ resource for wolves.

    Answer: limiting Difficulty: 2 Section: 1 Objective: 4

35. Members of a species compete indirectly for resources by competing for a(n) ____________________ and for social dominance.

    Answer: territory Difficulty: 2 Section: 1 Objective: 4

36. A population's ____________________ is the number of its members per unit area or per volume.

    Answer: density Difficulty: 2 Section: 1 Objective: 1

37. Deaths that are caused by a disease spreading through a population are ____________________ dependent.

    Answer: density Difficulty: 2 Section: 1 Objective: 4

38. A species' ____________________ includes that species' physical home, the environmental factors necessary for that species' survival, and all its interactions with other organisms.

    Answer: niche Difficulty: 2 Section: 2 Objective: 2

39. A type of interaction between two species in which both species are harmed is ____________________.

    Answer: competition Difficulty: 2 Section: 2 Objective: 3

40. Niche ____________________ is when each species uses less of the niche than it is capable of using, in order to reduce competition for resources with other species.

    Answer: restriction Difficulty: 2 Section: 2 Objective: 2

41. The organisms in a cow's stomach have a constant source of food; the organisms help the cow break down and use the grass it eats. This type of relationship is an example of ____________________.
Answer: mutualism Difficulty: 2 Section: 2 Objective: 3

42. A population's growth rate is its ____________________ rate minus its ____________________ rate.
Answer: birth, death Difficulty: 2 Section: 1 Objective: 2

43. The type of interaction between cats and mice is ____________________.
Answer: predation (or predator-prey)
Difficulty: 2 Section: 2 Objective: 3

44. A liver fluke is a(n) ____________________ that harms its host as it obtains food.
Answer: parasite Difficulty: 2 Section: 2 Objective: 3

45. A(n) ____________________ usually only weakens its host, while a(n) ____________________ usually kills its prey.
Answer: parasite, predator
Difficulty: 2 Section: 2 Objective: 4

46. A relationship in which two organisms live in close association, such as mutualism and commensalism, is called ____________________.
Answer: symbiosis Difficulty: 2 Section: 2 Objective: 5

47. If a pair of mice finds a place to live with plenty of food and no predators, the population of mice will probably undergo ____________________ growth.
Answer: exponential
Difficulty: 2 Section: 1 Objective: 2

48. Over a long period of time, two species can develop adaptations that increase the benefit of their relationship in the process of ____________________.
Answer: coevolution
Difficulty: 2 Section: 2 Objective: 5

49. A population has a(n) ____________________ growth rate when the death rate is higher than the birth rate.
Answer: negative Difficulty: 2 Section: 1 Objective: 2

SHORT ANSWER

50. The diagrams below show four different types of interactions between species. An arrow pointing from one organism to another means that the first organism has an effect on the second organism. Label each diagram with the correct type of interaction.

Answer:

a. mutualism
b. competition
c. commensalism
d. predation, parasitism

Difficulty: 3 Section: 2 Objective: 3

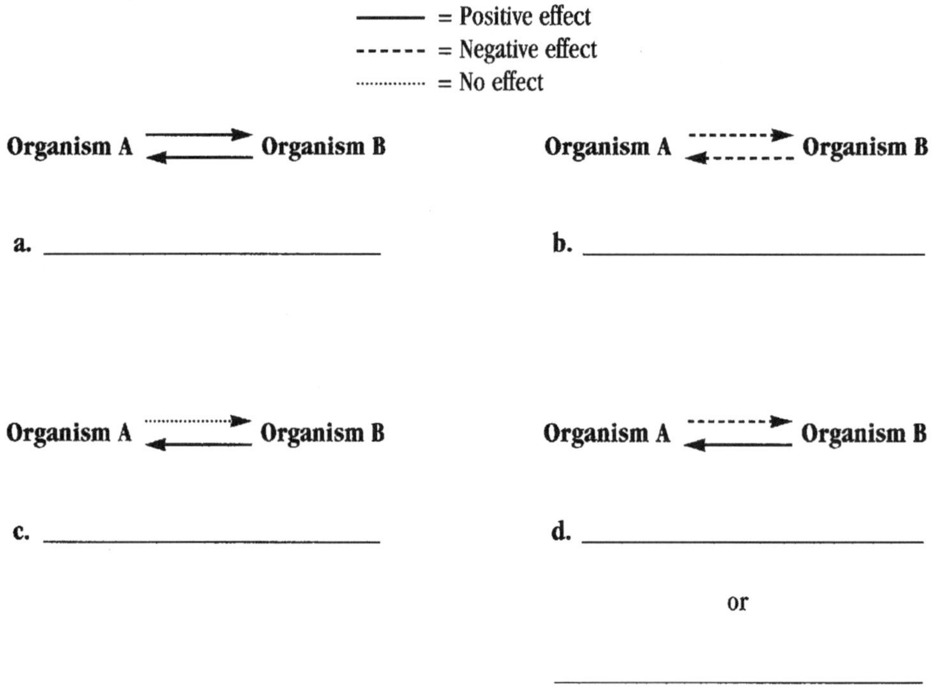

51. The cardon and organ-pipe are flowering cacti that depend on bats for pollination. The bats pollinate the cacti as they eat the nectar in cacti's flowers and spread seeds when they eat the cactus fruit. Studies of the cacti show that they are not producing as much fruit as they could. It was also noted that bats living near these cacti had been driven from their cave homes by local villagers. What is the relationship between the bats and the cacti? How did the reduction in the number of bats affect the cacti?

Answer:

The relationship between the bats and cacti is mutualism; the bats eat the nectar in cacti's flowers and spread pollen and seeds. With a decreased bat population, flowers will go unpollinated and fruit will go uneaten, reducing the number of opportunities for the cactus plants to reproduce.

Difficulty: 3 Section: 2 Objective: 3

52. Termites live almost exclusively on wood but cannot actually digest it themselves. Instead, they must depend on certain protozoa (single-celled organisms) that live in their gut to break down the wood into nutrients their body can use. In return, the termites provide an appropriate environment to sustain the protozoa. What is the relationship between the termites and the protozoa? How is this relationship similar to the one between humans and intestinal bacteria?

    Answer:

    The relationship between the termites and protozoa is mutualism—the termites receive nutrients in a usable form and the protozoa gain a hospitable environment. This relationship is similar to the one between humans and intestinal bacteria. Like the protozoa, the bacteria in human intestines break down some types of food that humans would otherwise be unable to digest. In return, humans provide a suitable environment for the bacteria, as do the termites for the protozoa.

    Difficulty: 3 Section: 2 Objective: 3

53. If a population of rabbits experiences exponential growth, what might happen to the population of coyote in the area. Explain your reasoning.

    Answer:

    Answers may vary. Sample answer: The population of coyote might also experience exponential growth. The coyote are predators of rabbits and would have an abundant source of food as the rabbit population grows at an increasingly faster rate. With plenty of food available, more coyotes would survive to reproduce.

    Difficulty: 3 Section: 2 Objective: 3

54. Predict what might happen to the population of rabbits and coyote if the rabbits exceed the carrying capacity of the environment. Explain your reasoning.

    Answer:

    Answers may vary. Sample answer: As the rabbits exceed the carrying capacity, they will run out of resources. Starvation and possibly diseases will severely reduce the population of the rabbits. The coyote will have less food available and their population will decline unless they find another food source.

    Difficulty: 3 Section: 2 Objective: 3

55. Choose any two species with a close relationship that might have coevolved adaptations and describe how the adaptations benefit both species.

    Answer:

    Answers may vary. Sample answer: A species of plant with red, tubular flowers and a species of hummingbird with long beaks may have coevolved. The hummingbird benefits by being attracted to the flowers where it finds an exclusive source of food. The plant benefits when the hummingbird pollinates the flowers it visits.

    Difficulty: 3 Section: 2 Objective: 5

56. Construct a table that compares and contrasts a parasite and a predator.

    Answer:

    Tables may vary, but should contain the idea that a parasite and predator are similar in that they benefit by obtaining resources while another species is harmed. They differ in that a parasite has a host that it lives in or on. Parasites may weaken their host but usually do not kill it to get the resource they need. Predators kill their prey to get food.

    Difficulty: 3 Section: 2 Objective: 4

57. Explain how two species can compete for the same resource even if they never come into contact with each other.

    Answer:

    Species can compete over time or space without meeting by utilizing the same resource at different times, such as one species of insect feeding on a plant during the day and another species of insect feeding on the same plant at night. Or, two species of plants that flower at the same time may be in competition for pollinators.

    Difficulty: 3 Section: 2 Objective: 3

58. Choose two populations and compare them in terms of size, density, and dispersion.

    Answer:

    Answers may vary. Sample answer: A population of Canada geese is often made up of a large number of individuals (size) in a small area such as a lake (density), while a population of redheaded woodpeckers is made up of a few individuals in each of several forested areas. The geese have a clumped dispersion and the woodpeckers have an even dispersion based on territories.

    Difficulty: 3 Section: 1 Objective: 1

59. List three reproductive behaviors of individual members of a species that affect the reproductive potential of a species. Which one of the three behaviors has the greatest effect on reproductive potential?

    Answer:

    producing more offspring at a time, reproducing more often, reproducing earlier in life.

    Difficulty: 3 Section: 1 Objective: 3

60. What are three density-dependent causes of death in a population? What are two density-independent causes of death?

    Answer:

    limited resources, predation, and disease; severe weather, natural disasters

    Difficulty: 3 Section: 1 Objective: 4

61. Choose an organism and give examples of parts of its niche. What is the difference between its niche and its habitat?

    Answer:

    Answers may vary. Sample answer: A niche includes a species' physical home, environmental factors for its survival, and its interaction with other organisms. A habitat describes the location where it lives. Parts of a pet dog's niche include its food and water, where it sleeps, other animals it chases, and its role as a family pet. The dog's habitat may be a residential neighborhood, as opposed to a forest.

    Difficulty: 3 Section: 2 Objective: 1

62. Give two examples of species that have the same habitat as hawks but different niches.

    Answer:

    Answers may vary. Sample answer: rabbits, snakes

    Difficulty: 3 Section: 2 Objective: 1

63. Aphids obtain the nutrients they need by sucking on the juices of host plants. This will later weaken the plants. What type of relationship do aphids and their host plants have? Explain your answer.

    Answer:

    Parasitism; the aphids feed on the juices without immediately killing the plant.

    Difficulty: 3 Section: 2 Objective: 3

PROBLEMS

64. Zebra mussels were accidentally imported to the Great Lakes from Europe in the 1980s. (They were stowaways on cargo ships.) These small mollusks have no natural enemies in the United States. Zebra mussels multiply quickly and attach themselves permanently to anything—fish, boats, rocks, pipes, buoys, or other zebra mussels! Huge water intake pipes for cities have been clogged, channel markers sunk, and marine engines damaged by the mussels. How could zebra mussels be eliminated from the Great Lakes?

    Answer:

    Answers may vary. Sample answer: The European practice of introducing the natural predators of zebra mussels into their habitat could be explored to see if it could be used effectively in the United States.

    Difficulty: 3 Section: 2 Objective: 3

65. Viruses are the cause of many infectious diseases, such as common colds, flu, and chickenpox. Viruses can be passed from one person to another in many different ways. Under what conditions do you think viral diseases will spread most rapidly between humans? What can be done to slow the spread of these viruses?

    Answer:

    Answers may vary. Sample answers: Viruses are density dependent and will spread most rapidly in crowded conditions. The spread of viruses could be slowed by having healthy people avoid crowded conditions and by having people with the viruses stay home.

    Difficulty: 3 Section: 2 Objective: 4

ESSAY

66. Imagine that one species no longer exists, or becomes extinct, immediately after the extinction of another species. Which relationship did the two species more likely have, competition or commensalism? Explain your reasoning.

    Answer:

    Commensalism; if the organisms were competing for resources, then the extinction of one would make more resources available for the other; thus, it should thrive. If species A derives benefit from species B and species B becomes extinct, then species A might also become extinct.

    Difficulty: 3 Section: 2 Objective: 3

67. Examine the graph below. Each line represents a different species. What type of interaction could be occurring between species A and B? Between B and C? Explain the reasoning behind each of your answers.

Answer:

The populations of species A and B fluctuate in an opposite (inverse) manner, suggesting competition in which one species monopolizes available resources. Alternatively, species A might prey on B (predation). Thus, as the number of predators (A) diminishes, the number of prey (B) would rise. Then as A increases consumption of B, the population size of A would increase and the size of B would diminish. The population sizes of B and C fluctuate together (in synchrony), which could indicate mutualism, or alternatively, a highly specialized type of parasitism or predation.

Difficulty: 3 Section: 2 Objective: 3

Population Size of A, B, and C Over Time

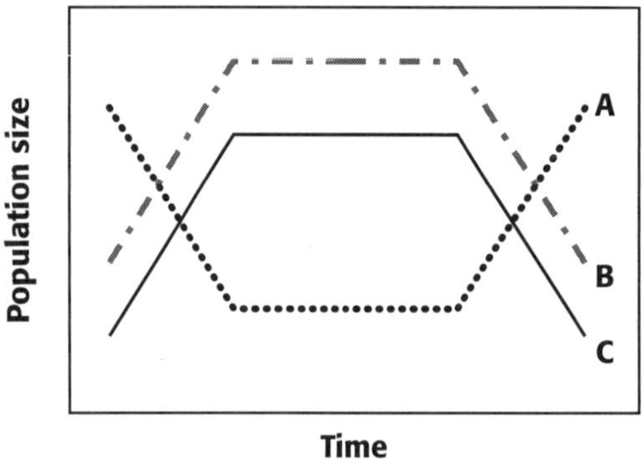

9997253272 2 3 4 5 6